The No Mammal MANIFESTO

DIET FOR A NEW, AND MORE SUSTAINABLE WORLD

ADAM ROGERS

No Mammal Manifesto:
Diet for a new, and more sustainable world

The author has done extensive research to be able to render the most recent and pertinent information available on this topic. However, this book is not meant to provide legal, medical, or any other professional advice on health-related affairs. The author and the publisher disclaim any personal liability, loss, or risk that may be incurred as a consequence of the use and application of this book, either directly or indirectly.

Phoenix DA
Østergade 19, 8900 Randers C, Denmark

Library of Congress Cataloging in Publication Data
Rogers, Adam, 1963 –
No Mammal Manifesto: *Diet for a new and more sustainable world*
Adam Rogers, 2019
Includes bibliographical references
Includes index
ISBN 978-87-971254-0-3
EAN 9788797125403
US$ 9.95, CAN$9.95, EUR9.95
Cover art and interior by Kerrie Robertson Illustration

Printed in USA and Denmark

TABLE OF CONTENTS

ACKNOWLEDGMENTS

OF THE MANY PEOPLE TO whom I owe a debt of gratitude for their input, resources and acting as sounding boards, I would like to recognize several in particular. Patrick Leung of New York and New Zealand, with whom I have had many inspiring conversations, and who graciously reviewed the early drafts of this manuscript. Gustavo González of the United Nations Food and Agricultural Organization responded to my many repeated requests for reports and references and provided much valuable input. Gillian Rogers did the final check on the manuscript and discussed with me many of the ideas therein before pen hit paper. Janet Mills of Amber-Allen Publishing has always provided a helpful reality check when I needed it most. My sisters, Rima and Lara Devitt, have remained committed to vegetarianism since they were teenagers, and continue to inspire me with their insights and cooking. Rima owns what is reputedly the best restaurant in Edmonton, Alberta: The Blue Plate Diner.

Others who peer reviewed the *Manifesto* before it went to press and to whom I owe a mountain of gratitude include Matt Mulford, my esteemed professor at the London School of Economics; Sid Kane, editor and communications officer formerly with the UN Development Program and the World Bank; Chris Fiscus, my editor at *The Lumberjack*; Judy Rae Hallcom, my publisher at Earth News; Tore Brevik, former communications director at the UN Environment Programme; and Cherie Hart, former UN communication advisor and author of one of my all-time favorite books, *From Hollywood to Holy Wars*. Linda Kleinschmidt edited the manuscript and made it more readable.

Thanks go also to Bekah Martin for letting me know about rennet, thus forever changing my view of cheese. I would also like to thank and recognize Frances Moore Lappé, who first set me on this journey many years ago with her book, *Diet for a Small Planet*. Lastly, and without hesitation, an expression of gratitude to my two tentmates in Wyoming's Wind River Range during the winter survival course at the National Outdoor Leadership School (NOLS). I have long since forgotten their names, but without their words and encouragement, I probably would not have started on the journey of inquiry that led to this *Manifesto*.

For Sage, Addison and Alena

INTRODUCTION: WHY THIS PARTICULAR MANIFESTO?

AS HUMANS, WE ARE ONE member of the mammal family. We have been walking around on this beautiful planet for around 200,000 years. Yet in that very limited amount of time we have had an enormous impact– mostly negative. Much of this destructive influence has been directed by what we choose to eat and how we go about securing our meals.

How we humans choose to feed ourselves should in theory nurture human health and support environmental sustainability. The current system of food production and distribution, however, is pushing our planet beyond the limits of what it can actually support.

A mammal-centered diet is not very healthy – either for us or our planet. Empirical studies have demonstrated that reducing or eliminating mammals from our diet can add years to our lives while also improving how we feel throughout those years. Furthermore, by eliminating mammal meat from our diets and embracing this *No Mammal Manifesto*, we will be contributing to solutions that will ensure that we have an ecologically-sound planet to enjoy in our new-found longer years of good health.

Empirical studies have demonstrated that reducing or eliminating mammals from our diet can add years to our lives.

The point of this book is when we have a choice, and most of us do have many choices on the menu, we should choose wisely – for our health, for the health of our planet and for the health of our children in whose hands we leave this planet when we die. Indeed we do need to eat to survive – but as omnivores we can choose to kill or not to kill, and if we kill, we can choose what, when and how.

I hope this book will encourage the reader to think differently about eating mammals, not just because of the mythical bond we share with them (Chapter 4), but because of an understanding that to eat mammals (when we don't have to) makes no sense from a variety of perspectives – including those related to our health (Chapter 1), the environment (Chapter 2) and the economy (Chapter 3).

A BIT OF BACKGROUND

The views within this *Manifesto* did not come to me in a sudden epiphany. I did not suddenly stop eating mammals because someone convinced me not to, nor because I was inspired by a sudden epiphany to change my diet. I evolved into doing so over several years, and the reasons why were complex and motivated by numerous factors, which I have presented in these pages. I wrote this book because I hope to convince you, the reader, to at least take some of my thoughts into consideration and begin (if you are not already doing so) to make conscious and fully informed decisions about the consequences, both personal and planetary, of your dietary decisions.

People whose dietary preferences are out of the mainstream are often asked to explain themselves – unless their restrictions are due to an allergy or a religious dictate. When Muslims turn down pork, no one asks them why they do so. When Hindus turns down a beef burger, they are not asked to explain why. When someone is lactose intolerant, he or she just needs to say “no, thank you” to a glass of milk, and no one will press for a reason.

If you are vegetarian, however, your dinner companions may either act defensive or be genuinely curious and ask you a plethora of questions -- from how long you have been a vegetarian to how “on God’s green earth” you ever do find anything enjoyable to eat.

The usual response I hear from vegetarians is that they wish to reduce animal suffering, benefit the environment, or lead a healthier life. All these ideas ring just as true for those of us who have chosen to exclude mammals from the menu, but the explanation in this instance becomes a bit more complicated.

People often will say, I "don't eat red meat," but that misses the point. I did not stop eating mammals because their meat is red. I stopped eating mammals *because* they are mammals. I did not stop eating meat that is red, I stopped eating meat that was clogging up my red arteries and putting me at an increased risk of heart disease and colorectal cancer. I do not avoid meat that is red. I eschew meat, the industrial production of which is destroying an earth that should always remain green.

I was not always this way. I have lived, worked, traveled, or visited most countries on this precious planet, and in many of them, people do eat a wide variety of things that move and must be caught, captured, or raised in captivity. As part of my travels, I usually joined my hosts in eating whatever they were eating -- pretty much everything from cats and dogs to snakes and even insects.

In the winter of 1980, I found myself trekking through Wyoming's Wind River Range with a backpack and a pair of old Army cross-country skis. I was 16 years old and taking a winter survival course at the National Outdoor Leadership School. I had two tentmates – a doctor and a psychologist. Both were vegetarians. As a smart-mouthed teenager from the Yukon, who thought it unnatural not to eat meat, I gave both of my tentmates a hard time and teased them relentlessly at every meal.

"But how can you comment on something you know absolutely nothing about," the doctor finally asked me. "When you don't know what you are talking about, you are all talk and no bite – just hot air with no substance."

After a few days of this debate, I finally agreed to stop eating meat for a few months. They both told me I would feel stronger, healthier, and more alive. I doubted them, but I accepted the challenge. The only specific advice they gave me was to read *Diet for a Small Planet*, the 1971 bestselling book by Frances Moore Lappé that outlined the environmental impacts of meat production. Lappé explained how world hunger was not caused by a lack of food, but rather by ineffective food policies and the industrial production of animal products.

The doctor said I needed to learn about balancing my amino acids. To be a vegetarian, he said, is to be fully aware of one's body and the effects different foods have on it – combining the right ingredients to get the maximum effect in terms of energy and well-being. If I just ate grains, for example, I would not get enough protein. However, by balancing grains with soy beans, I would have more than enough energy to climb a mountain. I learned this advice to be true years later when I hiked and climbed 300 kilometers around the Annapurna Range in the Himalayas, eating nothing but lentil beans, rice, and vegetables.

Unfortunately my tentmates never had the opportunity to say to me "I told you so." I lost touch with them, as I continued to explore the world and my place in it, paying particular attention to what and how people eat. Two years later, I had run out of new places to see in North America, and so I set off on a one-way ticket to Africa – and a journey that would take me through fifty countries in five years on a budget of less than $100 a month.

For most of this long journey, I stuck to a vegetarian diet although I did occasionally eat meat when my hosts offered it. When you travel on a shoestring budget, one naturally will rely on the hospitality and generosity of strangers. But when you are hungry, and your host kills a chicken or a goat to honor their invited guest, it would be poor manners to decline. Also, I learned that balancing one's amino-acids, vitamins, and protein can be challenging when traveling in remote areas of the world, so if given the chance, I did sometimes choose to eat fish or chicken when it was available.

IN THE BEGINNING...

Before going forward here, let's back up a bit – actually way back—to review where we humans fit in with the big picture. Life on Earth first emerged about 3.8 billion years ago, initially as single-celled prokaryotic cells, such as bacteria. Multicellular life then evolved over a billion years, and it has only been in the last 570 million years that the kind of life forms that we are familiar with today began to evolve, starting with arthropods and fish.

All animals (that includes us) can be divided into six broad families that range from the simplest (invertebrates) to the most complex (mammals).

Invertebrates are the creatures that we, as mammals, have the least in common with – and therefore, in my view, they are okay to eat. Invertebrates were the first animals to evolve as far back as a billion years ago, and as such, they are characterized by their lack of backbones and internal skeletons, as well as their relatively simple anatomy and behavior. Today, invertebrates account for a whopping 97 percent of all animal species. This widely varied group includes insects, worms, arthropods, sponges, mollusks, fish, and octopuses. Fish evolved from their invertebrate ancestors about 500 million years ago and have dominated the world's oceans, lakes, and rivers ever since.

Vertebrates are a large group that is distinguished by having a backbone or spinal column. This group includes amphibians, reptiles, fish, birds, and mammals.

Let's look first at amphibians. Amphibians are characterized by their semi-aquatic lifestyles which require them to stay near bodies of water to maintain the moisture of their skins and also lay their eggs and reproduce. Today they are among the most endangered animals on earth.

Reptiles, like amphibians, make up a fairly small percentage of terrestrial animals. This group can be divided into four basic categories: Crocodiles and alligators, turtles and tortoises, snakes, and lizards. Reptiles are characterized by their cold-blooded metabolism fueled by hanging out in the sun. Their scaly skin and their leathery eggs, unlike amphibians, let them survive away from lakes, rivers or streams.

Next are the fish. According to FishBase, a comprehensive online database on fish species, there are 33,100 species of fish in the world.[1] That is more than the combined total of all the other vertebrate species on Earth, including the mammals, amphibians, reptiles, and birds.

New research led by the American Museum of Natural History suggests there are about 18,000 bird species in the world[2], offering much more for our menu than mere mammals. Descended from dinosaurs, birds are characterized by their coats of feathers, their warm-blooded metabolisms, and their ability to adapt to a wide range of habitats – as seen with the ostriches of Australia and the penguins of Antarctica.

Let's move on to mammals, among the least diverse animal groups on earth and including only about 5,000 different animals. A myriad of mammal species once roamed the surface of this planet, creating a wonderful tapestry of life that covered every continent. Now, as a result of both the worst species extinction crisis since the demise of the dinosaurs and the rapid expansion of industrialized agriculture, 96 percent of the mammals present on earth today are livestock and the humans who eat them.[3] Just four percent are the lions, tigers, bears, elephants, giraffes, etc. – and our domesticated cats and dogs.

Mammals are characterized by their hair or fur (all mammals possess either during some stage of their life cycles), the milk they feed their young and their warm-blooded metabolism, which allows them to inhabit a wide range of habitats, ranging from the deserts of North Africa to the arctic tundra in Siberia.

The evolution of mammals has passed through many stages since the first appearance of our ancestors about 300 million years ago. The earliest mammals we know of were the egg-laying animals of the Prototheria subclass, which started out as something close to the platypus. However, we are not concerned with these animals for the purposes of this *Manifesto*, as there were no humans around then to eat them.

Mammals today range considerably in size from the 30–40 mm (1.2–1.6 in.) bumblebee bat to the 30-meter (98 foot) blue whale. All modern mammals give birth to live young, except for five species of egg-laying mammals (the platypus and four species of echidnas, all indigenous to Australia and New Guinea). However, like all mammals, the female monotremes nurse their young with milk.

Most mammal mothers, obviously including our own human mothers, have a placenta to facilitate the exchange of nutrients and waste between the blood of the fetus and that of the mother. This feature naturally creates a special bond between mother and child that other groups of species do not have. The largest groups of placental mammals are the rodents, bats and Soricomorpha (shrews and their allies). The next three biggest orders, depending on the biological classification used, are Primates (apes and monkeys), Cetartiodactyla (whales and even-toed ungulates), and Carnivora (cats, dogs, seals, and their allies). All mothers of the mammals one finds on the menu at McDonalds, Subway and Domino's Pizza have placentas in their wombs and mammary glands (breasts) to nurse their young.

OUR BRAINS

All mammals have a remarkably similar brain structure and nervous system. Of all the vertebrates, it is we mammals that have the biggest and most complex brain for our body size. On average, a mammal has a brain roughly twice as large as that of a bird of the same size, and ten times as large as that of a reptile of the same body size.

Size, although important (yes, when it comes to brains, size does matter), is not the only difference. There are also substantial differences in brain shapes. The hindbrain and midbrain of mammals are generally similar to those of other vertebrates, but dramatic differences appear in the front, which is greatly enlarged and altered in its structure as well.[4] This front part of our brain, called the cerebrum (which controls functions such as memory and learning), is much larger in mammals than it is in other vertebrates.

The cerebral cortex also strongly distinguishes the brains of humans and other mammals from the rest of the animals on Earth. The cerebral cortex is an outer layer of neural tissue, and it is separated into two cortices by the longitudinal fissure that divides the cerebrum into left and right hemispheres. In non-mammalian vertebrates, the surface of the cerebrum is lined with a comparatively simple three-layered structure called the pallium. In mammals, however, the pallium becomes a complex six-layered structure that scientists call the neocortex.[5] Several areas at the edge of the neocortex, including the hippocampus and amygdala, are also much more extensively developed in mammals than they are in other vertebrates.[6]

CONSCIOUSNESS

Consciousness is not an easy concept to define. It has been described as the state of being awake and aware of what is happening around you and of having a sense of self. The 17th century French philosopher, René Descartes, proposed the notion of "cogito ergo sum" ("I think, therefore I am"), which is the idea that the mere act of thinking about one's existence proves there is someone there to do the thinking. Descartes also believed the mind was separate from the material body — a concept known as mind-body duality. Yet if this concept were true, how could you knock a person unconscious by simply knocking him on the head (and cause him to lose some memory, if the knock was forceful enough)? There must be something in the head that makes consciousness possible. I believe that "something else" is what all mammals have in common.

When we discover and accept that all mammals have a consciousness as we do, we may come to the obvious realization that the way we treat and eat them should be reconsidered.

When mammals experience elevated levels of stress, their bodies release hormones that degrade the quality of the meat on their bones, producing what the industry calls "pale soft exudative" meat from pigs and "dark, firm, and dry" (DFD) meat from cows or sheep.

These stress hormones, such as adrenaline, cortisol, and other steroids, can lead to cardiac problems, impotency, and general fatigue in the humans that consume that meat.[7] This phenomenon is seen only in mammals -- perhaps because mammals have the consciousness to know they are about to be killed and to freak out about it.

Some people may argue that humans are the only mammals with consciousness and a conscience. It makes sense to me also that if we want to eat mammals, we would need to take this position, because if we became fully conscious of the consciousness in all mammals, we could not in good conscience eat them. I do believe this view will change over time, as we become more aware of and more respectful toward our fellow mammals. After all, it was not all that long ago when in my own country, the United States, those of European descent classified humans of African descent as being legally only three-fifths human.[8] It is now commonly accepted in America that all humans are in fact created equal. When we discover and accept that all mammals have a consciousness as we do, we may come to the obvious realization that the way we treat and eat them should be reconsidered.

CULTURAL AND RELIGIOUS TABOOS

There are individual cases where modern societies mull over the morality of eating our fellow mammals. In North America and Europe, most people I know would shudder at the thought of eating "man's best friend," the dog. However, we won't hesitate to chow down a hot dog, if it is made from the dog's close cousin, the pig – which according to most studies is much smarter and intuitive than a dog. Pigs have outperformed three-year-old human children on cognition tests and are smarter than any domestic animal. Overall, the research suggests that pigs have excellent long-term memories and are skilled at mazes and other tests that require them to locate specific objects. They have even been taught to put together simple puzzles – something a chicken or a turkey could never do.[9] My point? If you eat a hotdog, make sure it is made of chicken or turkey.

About a third of the planet's population -- who are adherents to both the Jewish and Islamic faiths – will not eat pork because of Scripture. But long before the emergence of the Old Testament and the Qur'an, people in the Middle East had largely cut pigs from the available menus. This decision was probably for economic and health reasons once chickens were domesticated. Chickens are a more efficient source of protein than pigs are and only require 900 gallons of water to produce two pounds of chicken, while it takes 1,500 gallons of water to produce two pounds of pork. Also, chickens produce eggs, an important secondary product that pigs do not offer. Chickens are much smaller and can be consumed within 24 hours of being killed, thereby eliminating the challenge of preserving large quantities of meat in a hot climate. Chickens are also easier to transport than pigs, and thus they were probably selected as the animal protein of choice by our semi-nomadic ancestors.

Indians also have a special mammal that is taboo for their dinner table. In India as well as other communities of Hinduism, such as Trinidad, Bali, and Fiji, killing and consuming cows is a sacrilege of the highest order.

Most Americans would likely gag at the thought of eating horse meat. However, horses are considered a delicacy in many countries around the world. Mexico, France, Switzerland, Kazakhstan, Belgium, Japan, Germany, Indonesia, Poland, and China are among the nations where you will find horse meat on the menu in many a fine restaurant.

In 2002, as the Republic of Korea prepared to host the World Cup, Brigitte Bardot, the French actress turned animal rights activist, led a global campaign against Korea's practice of eating dogs, calling it "barbaric." The International Federation of Football Associations (FIFA) received thousands of calls and letters condemning the practice. Korea fought back, led by the opposition in Parliament. Support was strongest among the patrons of the 6,000 restaurants that thrive on a mixture of dog stews, soups and *satays* washed down with alcoholic drinks flavored with pulverized cat. [10]

The amazing realization for me is that dogs somehow sense when humans eat other dogs. I know this through an experience I had many years ago. After dining on some roast canine in the remote Northern province of Benguet in the Philippines, nearly every dog I passed after that meal on the dusty street looked at me and growled. That was the first and the last time I ate dog meat.

Most Westerners are appalled by and sickened at the thought of eating dogs, and most Americans would also shudder at the thought of a horsemeat hamburger. One of the objectives of this book is to get the reader, if currently an omnivore, to consider broadening his or her circle of compassion to include other mammals.

THE FOOD CHAIN

What I clearly seek to dispel here are the common assumptions that we need to eat mammals to survive. For humans, eating meat is a choice, not a necessity. We are omnivores, not carnivores. It would be altogether a different matter if we found ourselves in a position where we need to kill something to survive or in a desperate scenario where the choice is either eat or be eaten. But when we have a choice, between killing a mammal or finding something else to eat, my goal is to present a compelling case for people to choose the latter. This book is my personal attempt to achieve that goal.

For humans, eating meat is a choice, not a necessity.

In ecology, a food chain is a series of organisms that eat one another, so that both energy and nutrients flow from one to the next. For example, if you have a steak for dinner, you might be part of a food chain that looks like the following: grass → cow → human. Each stage of this process is called a trophic level.

The primary producers include a group of organisms that are producing their own food. As with the given example, plants are primary producers based on their ability to manufacture their food through photosynthesis. At the next trophic level in a food chain or ecological pyramid, the organisms feed on the primary producers and are referred to as primary consumers, i.e. herbivores, such as cows and hippos, which eat the grass. Organisms that feed on the herbivores, called predators, occupy the next trophic level. At the top are the apex predators like the lions and tigers and bears.

Humans often believe they occupy a natural position at the top of this ecological pyramid. However, science does not support this assumption. In 2013, a group of French researchers calculated human trophic levels, and found that on a scale of 1 to 5, with 1 being the score of a primary producer (a plant) and 5 being a pure apex predator, they discovered that based on their diet, humans score only a 2.21—roughly equal to an anchovy or pig.

FIGURE 1: A SIMPLIFIED ECOLOGICAL PYRAMID

These researchers, led by Sylvain Bonhommeau of the French Research Institute for the Exploitation of the Sea, used data compiled by the United Nations Food and Agriculture Organization (UNFAO) to construct models of people's diets in different countries over time. They then used that data to calculate our human place in the food chain, using data from 176 countries from 1961 to

2009. Calculating the trophic level for humans was very straightforward: If a person's diet is made up of half plant products and half meat, his or her trophic level will be 2.5. More meat, and the score increases; more plants, and it decreases.

Their results confirmed common sense. as a species, we are primarily omnivorous, eating a mix of plants and animals, rather than being top-level predators that only consume meat.[11] The best evidence for this assertion is probably our teeth. Like carnivores, we have incisors for biting, tearing and ripping apart meat, as well as molars for chewing (as herbivores do). Animals with the same kind of diverse teeth tend to be omnivores.

In other words, we can eat meat, but we don't have to eat it. It is a *choice*. Also, when we do choose to eat meat, we can choose which type of meat we want to eat – mammal or other animals that do not include our close cousins. However, in many developed countries today, such as the United States, people are choosing to eat mammals every day – and quite often a few times a day.

THE FAST FOOD CHAIN

Nothing exemplifies the fixation on fast food in America more than the ubiquitous presence of the Golden Arches. The coordinates of N 45.45955 W 101.91356 identify a spot in South Dakota that has the peculiar distinction of being the "McFarthest point" in the lower 48 states.[12] The nearest McDonald's is 107 miles away from that spot, as the crow flies, or 145 miles away if traveling by car. It means that any hungry motorist is always within a full tank of gas of a McDonald's restaurant anywhere in the continental United States. That is just in one country – albeit a country that consumes a lot of hamburgers. Worldwide, as of early 2019, more than 69 million people eat at 36,899 McDonald's restaurants and consume "more than 75 hamburgers per second, of every minute, of every hour, of every day of the year," according to the company's *Operations and Training Manual.*[13]

Industrialized global meat production has transformed the food chain into a fast-food chain.

Because many of us have chosen to eat meat every day, industrialized global meat production has transformed the food chain into a fast-food chain. In the United States alone, the total mammal meat production (beef, veal, pork, lamb and mutton) totaled 52.1 billion pounds in 2017, three percent higher than the previous year. These figures come from the United States Department of Agriculture[14] but they do not include the actual number of mammals killed. The Humane Society of the United States on its website puts that total for 2015 at 28.7 million cows (an increase of 4.3 percent from the previous year), 115 million pigs, 2.2 million sheep and lambs – all raised and slaughtered within a single year in the United States alone. In Fiscal Year 2018, according to the North American Meat Institute website, there were 835 Federally-inspected livestock slaughter plants in the United States. An additional 3,773 plants process mammal meat, but they do not slaughter them.

In developing countries, according to the FAO, mammal meat consumption has been growing at five to six percent annually. This growing expansion of livestock production has severe environmental implications, for example, the expansion of land for livestock development. Livestock sector growth has been a prime reason for deforestation in many countries, such as Brazil and Argentina.

Global industrial mammal meat production has quadrupled in just the span of my lifetime, from 86 million tons in 1963 (the year I was born) to the current total of 340 million tons (308.4 trillion kg).[15] Yet the human population of the planet has only a bit more than doubled. This trend of meat production is projected to continue, especially as the growing urban middle classes in China and other emerging economies adopt the burgers-and-steaks diet of North America and Europe. According to the UNFAO, global meat production could increase to 501 million tons by 2050.

THE PET FACTOR

If all the cats and dogs currently living in the United States were together in their own country, they would rank 5th in global mammal meat consumption, behind only Russia, Brazil, the United States and China. In other words, cats and dogs in the U.S. eat more meat than all of the people and their pets in France and Germany. As outlined here in the chapter on environmental considerations, producing all this meat has an enormous impact on the environment. According to research on climate change conducted by Professor Gregory Okin at the University of California at Los Angeles (UCLA), total meat consumption by dogs and cats is responsible for the emission of 64 million tons of carbon dioxide a year – around the same climate impact as driving 13.6 million cars.[16]

This situation can be easily rectified with some effort. The cat is considered by scientists to be a strict carnivore while the dog is considered to be an omnivore. That means that cats eat meat – but it does not have to be mammal meat. Although we tend to think that our dogs need meat, they actually will eat everything. I further address the issue of what to feed our pets in the last chapter, Final Thoughts and Reasonable Recommendations.

The next chapters address the different arguments, or reasons, for reducing or avoiding mammal consumption, beginning with the first reason, which is your health. I then look at the impact that industrialized livestock production is having on the environment, before turning to the economic arguments that suggest we should stick to fish, chicken or vegetables.

REASON 1:

YOUR HEALTH

The **FIRST** *reason to reduce or eliminate mammal consumption from your diet is simple. Do it for your health.*

A MAMMAL-CENTERED DIET IS REALLY not very healthy. Empirical studies have demonstrated that reducing or eliminating mammals from our diet can add years to our lives, while also improving the way we feel throughout those years. Researchers at Oxford University estimate that by 2020, 2.4 million deaths annually will be attributable to the consumption of mammals — as well as a $285 billion health-care bill for those who cling to life in a hospital bed. The World Health Organization links these deaths to diabetes, heart problems, and cancer – all a result of eating beef, lamb and/or pork on a regular basis.

A January 2019 study by the EAT-Lancet Commission on Food, Planet, and Health, a collaboration between the EAT Foundation, *The Lancet*, Wellcome Trust, and the

Stockholm Resilience Centre, outlined the ideal healthy diet – one that is best for the health of the individual and the planet. Thirty-seven scientists from 16 countries (all international experts in health, nutrition, and sustainability), argued that despite the rapid rise in mammal meat consumption over the past several years, a significant proportion of people worldwide remain undernourished and overweight because their food lacks essential nutrients. This report found that these unhealthy diets posed a greater risk of death and disease than "unsafe sex, alcohol, drug, and tobacco use combined." [17]

Some nutritionists refer to the 'triple burden' of malnutrition weighing on most countries as "undernutrition," "micronutrient deficiencies," and "overweight and obesity." Different forms of malnutrition can also co-exist within the same country, the same household, and even for the same individual. In the United States alone, which boasts some of the highest mammal meat consumption in the world (after Argentina), malnutrition – particularly as micronutrient deficiency and obesity – is a reality for too many adults and children. Indeed 31 percent of Americans over the age of nine may be at risk of having a vitamin deficiency or anemia; more than 70 percent of U.S. adults over the age of 20 are obese or overweight[18]; and about 14 percent of kids under the age of five have weight problems.[19]

One of the reasons for this situation (although not the only reason) is an obsession with hamburgers. A research team from Boston University's Slone Epidemiology Center examined the association between the consumption of hamburgers from fast-food restaurants and obesity and found a direct correlation.[20] The link between obesity and certain deadly chronic diseases — arthritis, cardiovascular disease, hypertension, Type 2 diabetes, and some cancers, among others — is well documented.

Wartime history provides more important information regarding the correlation between animal-based diets and morbidity-mortality rates. As documented in the 2011 film, *Forks Over Knives*, during the 1930s, cardiovascular disease among Norwegians steadily increased. When the Germans occupied Norway in 1939, they confiscated all the livestock and farm animals to supply their own troops in their occupied territories. As a result, the Norwegians were forced to eat fish and plant-based foods. What was thought to be a curse, however, turned into a blessing as deaths from coronary artery disease plummeted during the war years from 1939-1945. When the Norwegians finally got their cows and pigs back at the end of the war, heart disease climbed once again.

Dr. Caldwell Esselstyn explains it this way in the film: "The native population subsisted on whole grains, legumes, vegetables, and fruit. Almost immediately death from heart attacks and strokes in Norway plummeted. With the cessation of hostilities in 1945 animal products became available as well as an immediate return to the pre-war levels of deaths from these illnesses. It is a powerful lesson in public health about the cause and cure of our most common killer--heart disease."

The proportion of men with low sperm concentration was three times higher for the sons of beef eaters than it was for men whose mothers did not eat beef.

There is even an emerging body of research that links decreased male sperm quality to mothers who eat beef while pregnant. A study in five US cities between 1999 and 2005 examined semen parameters in 387 partners of pregnant women in terms of the amount of beef their mothers reported eating while they were pregnant. The proportion of men with low sperm concentration was three times higher for the sons of beef eaters than it was for men whose mothers did not eat beef.[21] The researchers determined that the main cause for the low sperm counts was the anabolic steroids and other xenobiotics found in meat from industrialized livestock. The same problem would presumably not be found in the sons of mothers who eat only range-fed beef – although I have been unable to find a study that verifies this.

YOUR HEART

High saturated fat levels in mammal meat have long been known to contribute to heart disease, the leading cause of death in the United States today. Your body needs healthy fats for energy and other body functions, but too much saturated fat can cause cholesterol to build up in your arteries, thereby increasing your risk for heart disease and stroke.

Those who ate roughly four ounces of mammal a day were more likely to die of cancer or heart disease.

A joint study of more than a half a million older Americans conducted by the National Institutes of Health (NIH) and the American Association of Retired Persons (AARP) concluded that those who ate the most mammal meat over a ten-year period were likely to die sooner than those who ate smaller amounts. Those who ate roughly four ounces of mammal a day were more likely to die of cancer or heart disease than were those who ate the least, about half an ounce a day.[22]

A report from the Harvard School of Public Health also determined that regularly consuming mammal meat could lead to an untimely or early death.[23] Their data was taken from a study that followed more than 72,000 women for 18 years. They discovered that those who ate a Western-style diet high in red and processed meats had an increased risk of heart disease, cancer, and death. Another study by the same researchers followed 121,000 men and women for 24 years. All the participants submitted information about their diets every four years. Over the course of this study, almost 24,000 of the participants died. Death rates among those who ate the most red meat were higher than for those who ate the least. It found that people who ate one additional three-ounce serving of red meat daily faced a 13 percent higher risk of premature death. If that serving was processed meat (such as bacon or hot dogs), the risk went to 20 percent.[24]

One reason for the high incidence of heart disease for people who eat mammals is Trimethylamine N-oxide (TMAO), a dietary by-product that forms in the stomachs of those who eat mammal meat. Within just a couple of hours of consuming a steak, TMAO levels in the blood soar. This chemical, which has been definitively linked to heart disease, is derived in part from nutrients that are abundant in mammal meat. One of these, carnitine, is found in red meat and indeed derives its name from the Latin word *carnis,* the root for *carnivore.* Carnitine is not dangerous by itself, but when metabolized by bacteria in the intestines and thus ending up as TMAO in the blood, it becomes deadly.[25]

In one effort to understand this phenomenon further, 113 healthy men and women were enrolled in a clinical trial at the Lerner Research Institute of the Cleveland Clinic, a nonprofit academic medical center. All their meals were prepared for them, with dietary proteins coming from either mammal meat, white meat, or non-meat sources. Compared to the people eating diets rich in white meat or plant-based protein, those who ate a diet rich in mammal meat had three times the levels of this same TMAO chemical.[26]

The good news is that the TMAO increases were entirely reversible. When the participants of this study discontinued their mammal meat diet and ate either non-mammal meat or non-meat diet for another month, their TMAO levels decreased and were significantly below the levels associated with heart disease.

While high saturated fat levels in red meat have long been known to contribute to heart disease for people in general, a relatively new discovery suggests that a subgroup of the population may be at an even higher risk for a different reason -- a food allergen. Some of the emerging research from the University of Virginia Health System in Charlottesville is starting to provide an empirical link between beef and allergic responses in up to 20 percent of the population. The main allergen in red meat — a complex sugar called galactose-alpha-1,3-galactose (alpha-gal) — has been only recently identified and so most people do not even know they have that issue. Their allergy sets off an immunological chain reaction that leads to atherosclerosis or a build-up of fatty plaque in the arteries that hardens over time, narrowing the blood vessels, and eventually stopping the heart.[27]

INCREASED RISK OF STROKE

Stroke is caused by a blood clot or a burst blood vessel that stops blood flow to or in the brain. It is the third most common cause of death in the United States, killing about 800,000 people each year. Frequent consumption of mammal meat appears to increase the risk of being in this group. According to a study by researchers at the Harvard School of Public Health (HSPH) and the Cleveland Clinic, men who ate more than two red meat servings daily had a 28 percent higher stroke risk than did men who ate about one-third of a serving each day. People who ate chicken or turkey each day instead had a 13 percent reduced stroke risk over those who ate one daily serving of red meat.[28]

Men who ate more than two red meat servings daily had a 28 percent higher stroke risk.

The researchers who delivered these findings collected data from two large-scale health surveys that tracked more than 10,000 men and women for more than 20 years from middle age onwards. Not surprisingly, the investigators further found that substituting other proteins, such as nuts or fish, for one daily serving of mammal meat also reduced stroke risk significantly.

CANCER

Cancer is the second leading cause of death in the United States, accounting for nearly 600,000 deaths in 2015. [29] The World Health Organization has labeled mammal meat as a probable carcinogen, and processed mammal meat like hot dogs as a definite carcinogen – at the same level as cigarettes and asbestos. The analysis by the International Agency for Research on Cancer of the WHO found that eating 50 grams of processed mammal meat (about 4 strips of bacon or a single sausage) every day increased the risk of colorectal cancer by 18 percent.[30]

Eating only one hot dog a day raises the chances of developing colorectal cancer by 21 percent.

Another study conducted by the University of Hawaii goes even further, claiming that processed meats like hot dogs, ham, and sausage are responsible for a 67 percent increase in the risk of pancreatic cancer. The American Institute for Cancer Research claims that eating only one hot dog a day raises the chances of developing colorectal cancer by 21 percent.[31]

A crucial question when interpreting the results of any clinical trial is whether the size of the observed effect of a particular intervention (like eating processed meat) is clinically important. Several measures are used to quantify the "effect size" of an intervention – absolute risk reduction, relative risk reduction, and the number of cases observed. Scientists recognize that a clear understanding of these terms can help one make rational clinical decisions. While the studies noted here may report relative risk reduction, the fact remains that an increases in cancers were detected relative to the type of food that was consumed. Whether or not one can make a rational clinical decision to change one's diet based on that gathered information of course is always a personal choice. The information is here to inform your decision.

TYPE 2 DIABETES

It is well known that eating too much sugar and fat can increase your risk of getting Type 2 diabetes, but eating mammal meat can dramatically raise those chances as well. Conversely, several studies have concluded that choosing whole grains, nuts, low-fat dairy, fish, and poultry instead of mammal meat can significantly lower your risk of diabetes.

A 2012 study in the *Journal of the American Medical Association* examined the deaths of nearly 700,000 people in 2012 from heart disease, stroke and Type 2 diabetes. It found that nearly 50 percent of these deaths were related to poor nutritional choices. For those who already had diabetes, the risk of death increased if they consumed processed meats.[32]

Another study released in early 2017 from researchers in Finland analyzed the diets of more than 2,300 middle age men ages 42 to 60. At the outset, none of the participants had Type 2 diabetes –19 years later, 432 participants did. The Finnish study revealed that those who ate more animal protein and less plant protein had a 35 percent greater risk of getting diabetes. The researchers concluded that eating more plant and egg proteins could help prevent Type 2 diabetes.[33]

People who ate a single serving of red meat each day had a 19 percent higher risk of getting Type 2 diabetes.

Researchers at Harvard University were also curious about the links between eating mammal meat and diabetes. Their extensive research concluded that people who ate a single serving of red meat each day had a 19 percent higher risk of getting Type 2 diabetes than those who didn't. An even smaller size serving of processed red meat, such as one hot dog or two slices of bacon, increased the risk to 51 percent.[34]

THE B12 VITAMIN MYTH

Some of my friends who tenaciously adhere to a mammal-centered diet have often used B12 as their excuse. We do need Vitamin B12, as it is an essential nutrient that helps produce red blood cells and maintain the nervous system. A diet that lacks enough B12 can lead to deficiency symptoms that include anemia, constipation, loss of appetite, sore tongue, difficulty walking, numbness and tingling in the arms or legs, mood changes, disorientation, and memory loss.[35]

While some argue that red meat is the best source of vitamin B12, the vitamin is also found in other foods – for example, in eggs, milk, and chicken. An egg contains 0.89 micrograms of vitamin B-12 per 100 grams, or

15 percent of the daily required value (DV) of 6 micrograms. Two- percent milk contains 0.53 micrograms of vitamin B-12 per 100 grams, or nine percent of the DV, and 1.29 micrograms per cup, or 22 percent of the DV. Milk contains more vitamin B-12 in its recommended serving size than either eggs or chicken. Milk is also a good source of protein, calcium, phosphorus, potassium, and riboflavin, as well as vitamins A and D if it is fortified with these nutrients.

Chicken meat with skin contains 0.31 micrograms of vitamin B-12 per 100 grams, or five percent of the DV, and 0.21 micrograms per four-ounce serving, or four percent of the DV. Both eggs and milk are good sources of vitamin B-12. However, chicken provides other nutrients as well, which include protein, phosphorus, niacin, and vitamin B-6, so it is a good nutritious option to mammals

Other good sources of B12 that do not involve killing and consuming a fellow mammal include clams, mussels, crab, fish, fortified breakfast cereal, turkey, yogurt, and cheese.

THE GROWING RESISTANCE TO ANTIBIOTICS

The high volume of antibiotics used in industrial livestock production is mind boggling. Approximately 80 percent of the antibiotics sold in the United States are used for meat and poultry production. The vast majority is used on healthy animals to promote growth or prevent disease in mammals raised in crowded or unsanitary conditions.[36]

The livestock industry argues that while its widespread use of antibiotics may be causing antibiotic resistance on the farm, it still is not an important human health issue, and thus, little change in current practices is needed. Lawyers for the industry made their case in a June 2012 letter to Congresswoman Louise Slaughter of the House Rules Committee, which was cosigned by the American Farm Bureau Federation, the American Feed Industry Association, the American Meat Institute, the Animal Health Institute, the National Cattlemen's Beef Association, the National Meat Association, the National Pork Producers Council, and others. In the letter these groups declared that the use of antibiotics in livestock "contributes little, if anything, to the burden of human antibiotic resistance..."[37]

Numerous studies however have demonstrated that farm mammals that are raised for slaughter do in fact produce bacteria (i.e., Salmonella, Campylobacter, and Escherichia coli) that can be transferred to people. That transmission can occur from direct human contact with animals and also via food produced from animals with that resistant bacteria. Further, antibiotic-resistant bacteria can travel away from the farm in water runoff that carries manure and in airborne particles carried by the wind from the farm to nearby communities.[38]

These bacteria are frequently antimicrobial-resistant and can contaminate our food supply from farm to fork. In 2010, the U.S. Food and Drug Administration, the U.S. Department of Agriculture, and the Centers for Disease Control and Prevention all testified before Congress that there is a connection between the routine use of antibiotics for meat production and the declining effectiveness of antibiotics given to people.[39] Antimicrobial-resistant (AMR) infections in humans can produce longer illnesses, increased frequency of hospitalization, and treatment failures that can actually result in death.

Some of the types of bacteria that cause serious infections in humans have already developed resistance to most or all of the available treatments. Thus, we are running out of treatment options for these types of infections. The World Health Organization is recommending an overall reduction in the use of antibiotics in livestock to help preserve the effectiveness of these drugs for human medicine use.[40] My own recommendation, which would completely solve the AMR threat, is to reduce or completely eliminate the industrial production of meat for human consumption.

MAD COW AND OTHER ILLNESSES

Mammals also have a significant reservoir of diseases that can affect humans. While 60 percent of human diseases can be shared with other mammals, 75 percent of new infectious human diseases were first found in mammals. [41]

One such example is mad cow disease, or bovine spongiform encephalopathy (BSE), a transmissible, slowly progressive, degenerative, and fatal disease that affects the central nervous systems of adult cattle. A human version of mad cow disease, a variant called Creutzfeldt-Jakob disease, can come from eating a cow that is infected with BSE.

The worst recorded case of mad cow disease was in the United Kingdom in the mid-1980s when the cattle were believed to have been infected by being fed meat-and-bone meal that contained the remains of other cattle and certain sheep products. The cattle industry forced these herbivores to become carnivorous, and the results were horrific.

This outbreak spread throughout the UK due to this practice of feeding the young calves of dairy cows the slaughtered remains of their close relatives. From 1986 to 2015, more than 184,000 cattle were diagnosed, with the peak of new cases occurring in 1993. A few thousand additional cases were reported in other regions of the world. A few million cattle with the condition also likely entered the food supply during the outbreak.[42] To deal with the crisis, the British had to kill and burn about 4.4 million cows.

INDUSTRY RESPONSE

The meat industry wants you to believe that their product is good for your health. This view comes as no surprise, as the meat industry broadly is responsible for 5.4 million jobs and $257 billion in wages in the United States alone. An estimated 527,019 people have jobs in production and packing, importing operations, sales, packaging, and the direct distribution of meat and poultry products.[43] This enormous industry has to do everything it can to sustain and grow its profits. That is the nature of business, and also their right within the law.

The meat industry wants you to believe that their product is good for your health.

When the World Health Organization released its findings that meat causes cancer, the industry hit back hard, questioning the credibility of the researchers and their results. The leaders of prominent processing companies throughout the industry defended the meat industry. A similar study from the World Cancer Research Fund that linked processed meat to cancer received similar rebukes. Barry Carpenter, President and CEO of the North American Meat Institute (NAMI) even went so far as to declare that there was no significant relationship between meat and colon cancer and that the opposite was true – indeed that in some cases, there was a "protective effect."[44] On the NAMI website it actually states that we should be eating more meat. NAMI pegged the daily average meat consumption for Americans at 4.8 ounces for men, and 3.13 ounces for women – with a recommendation to increase that amount to 5.7 ounces for both.[45]

In response to the January 2019 EAT/Lancet report calling for an immediate reduction in mammal meat consumption to improve human health, the National Pork Producers Council issued a statement declaring the claim as radical, dubious, and even irresponsible. The National Cattlemen's Beef Association issued a statement, calling beef "nourishing and sustainable." Kay Johnson Smith, President and CEO of the Animal Agriculture Alliance, was quoted as saying "the Commission's radical recommendations to drastically limit meat and dairy consumption would have serious, negative consequences for the health of people and the planet."[46]

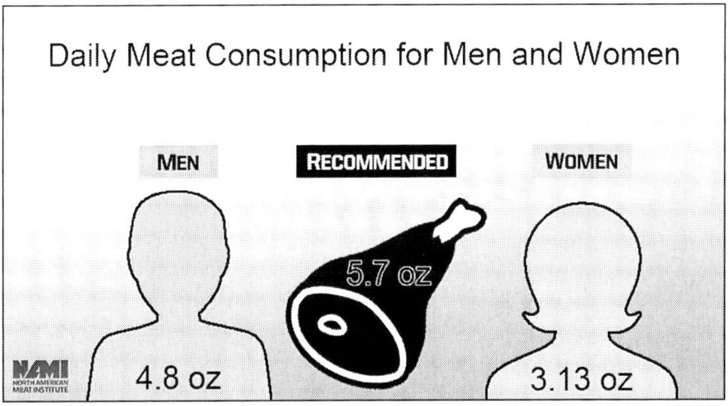

FIGURE 2: DAILY MEAT CONSUMPTION FOR MEN AND WOMEN (SOURCE: NAMI WEBSITE)

These statements on behalf of the meat industry are not surprising, given that the WHO and the EAT/Lancet recommendations, if implemented, would have serious, negative consequences on this industry's profit margins – something they would obviously prefer to avoid at all costs.

JUST TRY IT

Many people may respond to this chapter by saying the jury is still out on the effects of meat consumption on human health. Indeed, the battle lines have been drawn, and the meat industry continues to fight consumer advocates on the actual details. However -- you, as an individual, ultimately are the best judge of your own health (along with your doctor). If this chapter is not convincing enough, I would challenge you to adopt the same tactic I did in response to my two tent mates back in Wyoming: *Just try it.*

Giving up eating mammals can be extremely gratifying for your body and mind — and not just because you will be reducing your carbon footprint. It also has proven health benefits that include disease prevention and improved digestive health. However, still plan carefully to make sure you get all the nutrients your body needs. Start with a full medical checkup to establish a baseline – stress test, cholesterol levels, body fat percentage, etc. Then, reduce the amount of mammal meat you consume by 50 percent. Better yet, go *cold turkey* (literally, avoid the cow and eat the turkey). After three to six months, take a long walk and ask yourself how you feel. Do you feel stronger, lighter, and more energetic? Chances are you will feel all of the above. But do get another medical check up to compare the numbers. Upon reflection, you may embrace the *No Mammal Manifesto*, with gusto. You may even take it further still and cut out all types of meat. The point is to find what works for you and to balance that with what also works for our planet – something I will cover in the next chapter.

REASON 2: ENVIRONMENTAL CONSEQUENCES

The **SECOND** *reason to reduce or eliminate mammal consumption from your diet is because you care about the environment, and you depend on it for your own survival.*

ACCORDING TO AN EXTENSIVE STUDY by the Union of Concerned Scientists, the production and consumption of mammal meat is the second most environmentally harmful consumer activity in the world today. The only other human activity that is worse for the planet is our reliance on fossil fuels and the internal combustion engine to move us and our stuff around. This finding means that if you eat a hamburger in Alaska that was made from a cow raised in Brazil, you will be making a massive contribution toward healing the planet by re-considering your meal choice.

The preceding chapter looked at the personal health consequences of a mammal-centered diet, and was an argument for personal responsibility – like seatbelts or motorcycle helmets. This chapter is broader in the sense that it looks at how your personal food choices can affect everyone else.

The industrial production and consumption of livestock is undermining the very ecosystems on which you and I and 7.6 billion other people depend.

According to *The Consumer's Guide to Effective Environmental Choices*, one pound of beef on the table creates 17 times more water pollution and 20 times more habitat alteration than preparing its caloric equivalent in pasta.[47] In the Guide, Michael Brower and Warren Leon lay out 11 "priority actions, " that urge people to consider, for example, the cars they drive, the appliances in their homes, and the amount of meat in their diets. Brower and Leon identified what they consider to be the top four environmental problems of our modern Western lifestyles. These are air pollution, global warming, habitat alteration, and water pollution. Then they divided our household activities into broad and narrow categories and ranked them according to their contributions to these big four environmental problems. One of the biggest personal contributions you can make towards healing the planet, they asserted, aside from trading in your car for a bicycle, is to stop eating mammals.

The industrial production and consumption of livestock is undermining the very ecosystems on which you and I and 7.6 billion other people depend. For the past several millennia, we have enjoyed a period of relative climatic stability that has allowed humans to settle, farm, and create civilizations. According to the International Union of Geological Sciences (IUGS), the professional organization in charge of defining Earth's time scale, this recent period is known as the Holocene ("entirely recent") epoch. It began 11,700 years ago after the last major ice age.

Many scientists are now speculating that we are leaving the Holocene period and entering the "Anthropocen" (from the word *anthropo*, for "man," and *cene*, for "new" – a new global environment caused by human activity). The 2017 book by John W. Kress and Jeffrey K. Stine, entitled *Living in the Anthropocene: Earth in the Age of Humans* takes a vital look at this new era. The authors write that the root causes of the Anthropocene Age are the spread of agriculture, pollution, and urbanization. As we will see here, a heavy reliance on mammal meat consumption is one of the primary reasons for the unsustainable spread of agriculture.

In October 2018, scientists from around the world warned that we needed to dramatically reduce the amount of mammal meat we eat or face apocalyptic consequences. Beef consumption, in particular, needs to drop by 90 percent, and pork consumption needs to drop by about 80 percent if we are to restore ecological balance and increase our long-term chances of survival.[48] The difference in this percentage is linked to the efficiency ratios of converting grain to flesh (beef requires more resources to produce than does pork). The research, which was led by the University of Oxford, is the most complete to date, combining data from every country to assess the overall impact of food production on the global environment.

Eating two pounds of beef is responsible for more greenhouse gas emissions and pollution than driving around for three hours while leaving all the lights on back home.

However, despite the urgent appeals to reduce mammal meat consumption, the trend is still moving in the opposite direction. As more countries develop, much of the world is adopting American and European standards of living with an accompanying fixation on eating mammals. In the United States, each person now eats about 260 pounds of meat per year, while the average Brit consumes about 170 pounds.[49]

Fueled by rising incomes, mammal meat consumption in China grew sevenfold over the last three decades and a half. In the early 1980s, when there were fewer than one billion Chinese, the average person ate around

30 pounds of meat per year. Today, with an additional 380 million people, it's nearly 140 pounds per person, per year. With its higher population, the country consumes twice as much mammal meat as the United States -- 28 percent of the world's total. And the figure is only set to increase.[50]

Most of Africa and South Asia consumes less than 44 pounds of mammal meat a year. In all likelihood, at the current growth rates, worldwide mammal meat consumption is likely to double by 2050, according to sources at the UNFAO.[51] The planet simply cannot support the industrial production of that much meat, unless there are radical solutions discovered and implemented. The final chapter offers some possibilities for those solutions but one of the most obvious is to reduce, and advocate for the reduction of mammal meat production and consumption.

CLIMATE CHANGE

The Earth would be a lifeless rock like the moon if it were not for the thin layer of atmosphere above us that traps solar energy and insulates the Earth's surface from frozen outer space. The way the atmosphere traps solar energy is called the Greenhouse Effect, because the effect is just like a greenhouse or a closed car that heats up in the sun. For our planet, the clear windows that prevent the heat from escaping into space are made up of certain gases in the atmosphere, mainly water vapor, carbon dioxide, and methane. This delicate balance of celestial sun rays and carbon dioxide (or CO_2) is what makes our Earth such a nice place to live.

Since the dawn of history, the amount of CO_2 in the atmosphere was stable at about 280 parts per million (ppm) -- meaning that out of every million molecules in the air, 280 of them were carbon dioxide. It's a pretty small fraction of the atmosphere (0.028 percent), but it's the right amount of carbon dioxide to absorb just enough heat, so that the planet can maintain the overall average temperature that we and everything else on Earth have gotten used to living and thriving in.

The National Oceanic and Atmospheric Administration (NOAA) is an American scientific agency within the United States Department of Commerce that focuses on the conditions of the oceans, major waterways, and the atmosphere. According to NOAA, the level of CO_2 in the atmosphere is changing and escalating quickly. These scientists say global average atmospheric carbon dioxide in 2017 was 405 parts per million, with a range of uncertainty of plus or minus 0.1 ppm.[52]

The last time the atmospheric CO_2 amounts were this high was more than three million years ago when the temperature was about 3°C (5.4°F) higher than during the pre-industrial era, and sea level was 15–25 meters (50–80 feet) higher than it is today. Eighty feet of difference in sea level would wipe out most of today's coastal cities, turning places like Manhattan into Atlantis.

Scientists believe global temperatures will increase even further, continuing to rise for decades to come, largely due to the greenhouse gases produced by human activities that have yet to reach the upper levels of the atmosphere. The Intergovernmental Panel on Climate

Change (IPCC), which includes more than 1,300 scientists from the United States and countries around the world, has forecast an average temperature rise of 2.5 to 10 degrees Fahrenheit over the next century.[53]

SO WHAT'S YOUR STEAK GOT TO DO WITH IT?

While energy generation, transport, and construction are identified as the usual targets when governments seek to reduce emissions, the impact from food production has been somewhat overlooked. However, based on the current trend, with intensive agriculture increasingly geared toward livestock production, food production is now also a major factor to be considered.

Throughout much of the world, forests have been cleared to make way for livestock. The inefficient farming of cattle feed, together with methane emissions from cows and fertilizer use, creates as much greenhouse gas emissions as all the world's cars, trucks and airplanes combined.[54] Producing a kilogram of beef (two pounds) generates around 26 kilograms (57 pounds) of carbon dioxide, the highest of all the 197 foods examined using the U.S. Department of Agriculture's (USDA) food availability data set and a literature meta-analysis of emission factors for various food types.[55]

Eating a kilogram of beef is responsible for more greenhouse gas emissions and pollution than driving around for three hours while leaving all the lights on back home, according to Akifumi Ogino of the National

Institute of Livestock and Grassland Science in Tsukuba, Japan. Ogino and his team looked at calf production and focused on animal management and the effects of producing and transporting feed.[56]

Comprehensive research, led by scientists at the Oxford Martin School, found that shifting to a mostly vegetarian diet or even cutting down on meat consumption to within accepted health guidelines, would reduce greenhouse gases significantly.[57] A 2013 report from the FAO revealed that 14.5 percent of all human-induced emissions come from eating mammals. The report, *Tackling climate change through livestock,* says beef and cattle milk production account for most emissions, contributing 41 percent and 19 percent of the sector's emissions, respectively. Pig meat production is second, contributing nine percent to the sector's emissions [58].

According to the FAO study, the main sources of emissions are feed production and processing (45 percent of the total – with nine percent attributable to the expansion of pasture and feed crops into forests); fermentation from ruminants (39 percent); and manure decomposition (10 percent). The remainder of the carbon emissions from meat production is attributable to the processing and transportation of meat itself.

The report further states that the livestock sector can indeed make an important contribution to international efforts to curb climate change by voluntarily offsetting some of the sector's emission increases, since the worldwide demand for livestock products is expected to grow by 70 percent by 2050[59].

The beef industry, to its credit, is working to reduce its carbon footprint. The National Cattlemen's Beef Association, a trade group for beef producers, claims that improvements in cattle genetics and production methods have resulted in one-third fewer cows needed in 2015 to produce the same amount of beef produced in 1975.[60]

DISAPPEARING FORESTS AND LAND DEGRADATION

The quality of our land, sea, and water is deteriorating so rapidly that our ecosystems may soon be unable to sustain life as we know it today. The livestock industry already has degraded 20 percent of all the world's pastureland through overgrazing. It is also overusing our limited freshwater resources and has created vast oceanic dead zones from the nitrogen and phosphorus run-off that comes from chemical fertilizers.[61]

Just four commodities—beef, soy, palm oil, and wood products—drive most tropical deforestation. Of these four, beef has by far the largest impact.

The loss of primary tropical rainforest — the wildest and most diverse swaths — has increased as much as 25 percent since the 1990s. We are losing upwards of 80,000 acres of tropical rainforest daily, and significantly degrading an additional 80,000 acres every day. As the trees disappear, so do some 135 plant, animal, and insect species every day —some 50,000 species each year. Cattle ranching is cited as one of the primary reasons for the clearing of the forests – both for the cattle themselves, and to grow all the crops to feed them.[62]

Just four commodities—beef, soy, palm oil, and wood products—drive most tropical deforestation. Of these four, beef has by far the largest impact. Converting forest to pasture for beef cattle, largely happening in Latin America, is responsible for destroying 2.71 million hectares of tropical forest each year—an area about the size of the state of Massachusetts.[63]

The drive behind the incessant clearing rainforests is both growing demand for beef, as well as because much of the grazing land is rendered useless after a few years (maximum ten). The land suffers substantial losses of soil fertility and soil erosion because soil nutrients are rapidly depleted after clearing and grasses are soon replaced by less useful vegetation, causing farmers to clear yet more rainforest to feed and pasture their cattle.[64]

Data from the Brazilian Beef Exporters Association show that beef exports from that country increased 20 percent in 2017 to 132,000 metric tons and then an additional 11 percent in 2018 to 178,000 metric tons.[65] How many Brazilian cows is required to produce 178,000 tons of beef? A steer produces a 750-pound carcass after the fat and muscle are trimmed away. Remove the bones, and you get around 490 pounds of boneless trimmed beef. The 178,000 metric tons of beef exports in 2018 translates into 392,422,827 pounds of meat. Divide that figure by 490, and we get around 800,862 cows that are slaughtered each year in Brazil alone. Consider the amount of land it takes to produce all those cows each year, and you will start to understand what is happening to the rainforest.

IMPACT ON SCARCE WATER RESOURCES

Our mammal-centered diet is also delivering a devastating double whammy to our scarce freshwater resources. First, the enormous amount of water required to grow all the crops we feed to cows and pigs is staggering. A January 2012 report in *National Geographic* pointed out that irrigating the land for cattle feed uses almost three times as much water as for all the other foods combined. On the other hand, dairy cows require much less water, and their products (primarily milk and cheese) contribute the most calories to U.S. diets, but do not involve killing the cow.[66]

The enormous amount of water required to grow all the crops we feed to cows and pigs is staggering.

Just 55 percent of the world's crop calories are actually eaten directly by people. Of this amount, 36 percent is used for animal feed, with the remaining nine percent going to biofuels and other industrial uses. Recent studies find that global crop demands will likely increase by 60–120 percent by the year 2050 (using the baseline year 2005).[67] According to a 2011 analysis, 75 percent of all agricultural land (including crop and pasture land) is dedicated to animal production – and by far the lion's share of that amount is used to produce feed mammal meat.[68]

It is becoming more and more difficult to grow a sufficient amount of crops in the now increasingly depleted soils to feed all the cows and pigs. Farmers are thus turning to nitrogen-rich fertilizers to grow their crops. The chemicals in these fertilizers are percolating down into our freshwater aquifers and running downstream into our oceans. The result is that algae blooms are sucking all the oxygen from the water, and killing all life. And these "dead zones" are expanding like a giant plague: the U.S. National Oceanic and Atmospheric Administration (NOAA) announced in August 2017 that the dead zone in the Gulf of Mexico that runs along the United States coastline was the largest ever recorded, measuring 8,776 square miles – about the size of the state of New Hampshire or Massachusetts.

OTHER REASONS FISH ARE DISAPPEARING

The United Nations recently reported that nearly 90 percent of the world's marine fish stock has now been fully exploited, overexploited, or depleted.[69] Because of this overharvesting of fish, all 17 of the world's major fishing areas have reached or exceeded their natural limits, and fishermen the world over are increasingly heading back to port regularly with empty nets.

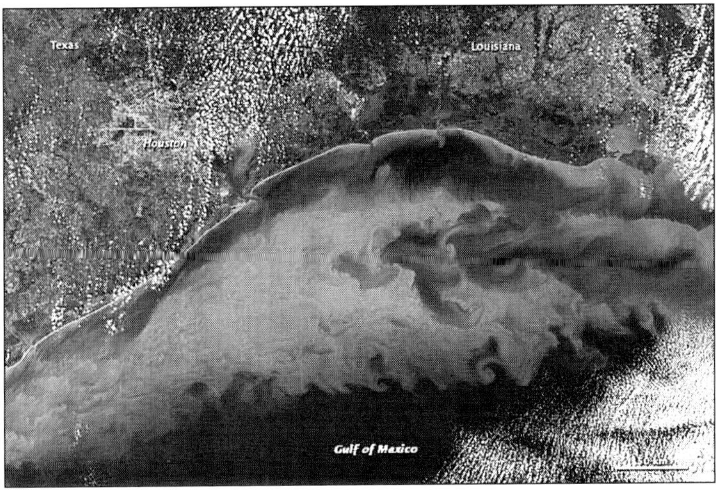

Satellite image of the Gulf of Mexico in 2017. Light colored water is the often nutrient-rich sediment flowing into the deeper ocean water. These nutrients facilitate growth of phytoplankton blooms, which lead to hypoxic conditions. Source: NASA

You might assume that all these fish are disappearing because we humans are eating them. I was surprised to learn that more than a third of all fish caught in the sea are fed to livestock or fed to farmed fish as a form of aqua-cultural cannibalism. [70] Thus, the livestock industry is killing sea life *en masse* both through nitrogen runoff from inefficient farming to grow feed crops as well as through overfishing to feed hungry cows and farmed fish.

More than a third of all fish caught in the sea are fed to livestock.

According to Sustainable Fisheries, which is a group of fishery scientists at the University of Washington, about 33 percent of the world's fish stocks are being exploited beyond reasonable sustainability.[71] That's right. A third of the fish that are left are still being overfished. Since studies show that 31.5 million tons, or 37 percent of all fish taken from the world's oceans each year is used to feed livestock, we could easily restore the balance simply by stopping the practice of feeding fish to cows – or by eating fewer cows.

MOVING IN THE RIGHT DIRECTION

As previously noted, while the people in emerging market countries like China are eating on average more mammal meat than before, other countries are slowly starting to wake up and realize the errors of their ways. In the United States, according to a 2017 survey conducted by the Natural Resources Defense Council, Americans reduced their beef consumption by 19 percent from 2005 to 2014. These positive changes in the overall American diet reduced emissions by the equivalent of the pollution from 57 million cars — despite a population growth of about 9 percent.[72]

Consumption of all meat in the United States, taken together, has been falling from a peak in 2007, according to the U.S. Agriculture Department. Per capita consumption of beef topped out in 1976, but remained America's favorite meat until the mid-1990s, when it was surpassed by chicken.[73]

Mintel, a consumer research firm, surveyed consumers to see why Americans are eating less beef. Of this group, 37 percent cited the cost of beef as the No. 1 reason. These respondents said they were eating more kinds of protein like chicken or tofu because of their lower cost. More than a quarter of the group ascribed the change to their concern about cholesterol and saturated fats. A separate study surveyed adults who had totally stopped eating meat. Of those surveyed, 52 percent cited taste as the top reason for eating plant-based proteins, outranking the concerns over diet (10 percent), animal protection (11 percent), the environment (13 percent) and health (39 percent).[74]

Another study conducted by researchers at the Johns Hopkins Center for a Livable Future collected survey responses from a nationally representative sample of 1,112 U.S. adults age 18 and older. Their study found that 41 percent of Americans are reducing their consumption of mammal meat, while 55 percent no longer are eating processed meats. Of those who reported eating less mammal and processed meat, 37 percent indicated they increased their consumption of non-mammal meat, such as poultry or seafood.[75]

Whatever the reason Americans are eating fewer mammals, it is clear that while their health is improving from their choice, the rest of us are benefiting as well by having a cleaner more stable environment. Let us all hope this positive momentum continues.

REASON 3:

ECONOMIC CONSIDERATIONS

The **THIRD** *reason for reducing or eliminating mammal consumption from your diet is because it is economically wasteful and unsustainable.*

THIS SECTION ADDRESSES THE *ECONOMIC considerations* of production, distribution, and consumption of goods and services – the inputs of grain, pesticides, labor, water, land, animals, etc., that go into producing mammal meat for public consumption.

If the negative externalities of industrial mammal meat production were included in its price, then meat would be very, very expensive, indeed a luxury item to be consumed as a special treat and rarely, if at all. The problem is that the true cost is NOT reflected in the price. Yet as Charles Portis reminds us in his book, *True Grit*, "You must pay for everything in this world one way and another."

The industrial conversion of calories from plants to animals – especially mammals – does not make for sound economic policy.

The industrial conversion of calories from plants to animals – especially mammals – does not make for sound economic policy in an increasingly overcrowded world. It is incredibly inefficient. Mammal production causes an enormous loss of calories that are grown in fields, since cereals and oil seeds must be cultivated to feed these animals. According to the United Nations Environment Programme (UNEP), the calories that are lost by feeding cereals to animals, instead of being used directly as human food, could likely feed an extra 3.5 billion people every year.

Furthermore, mammal meat is an inefficient source of calories. Nearly 60 percent of the world's agricultural land is now used for beef production, and yet beef accounts for less than two percent of the calories consumed worldwide. Beef makes up 24 percent of the world's meat consumption, and yet it requires 7.4 billion acres of land to produce it. In the United States alone, 70 percent of agricultural production goes to feeding livestock. In South America and Central America, five million acres of rainforest are felled each year to create cattle pasture. [76] All these economic inputs are undertaken to produce just 17 percent of the calories we consume!

According to the environmental organization, EarthSave, it takes 2,500 gallons of water, 12 pounds of grain, 35 pounds of topsoil, and the energy equivalent of a single gallon of gasoline to produce just one pound of feedlot beef. The conservative British newsmagazine, *The Economist*, puts the figure for lamb at six pounds of grain to one pound of lamb and "for beef starts at five to one and goes as high as 20 to one," depending on how and where the cow is raised. On the other hand, the ratio for chicken is only two to one. In other words, it takes two pounds of feed to produce one pound of chicken.[77]

A *National Geographic* calculation claims it takes 660 gallons of water to make just one hamburger.[78] That's enough water for three ten-minute hot showers. If you are trying or need to conserve water, your significant other may prefer you skip that burger rather than having to miss three showers.

Nitrates from cattle farming now contaminate the public water supplies of almost 1,700 communities across the United States.

The cattle industry in the United States is, to its credit, bringing this ratio down. An article in *Mother Jones* highlights several things that ranchers are now feeding their cows to move away from an outrageously-high dependency on corn, soy, and hay. Sawdust is a big winner, after removing the lignin, which makes it hard to digest. The lignin is removed by soaking the sawdust in nitric acid. Other things that are now fed to cows include surplus gummy bears, marshmallows, hard candy, sprinkles, chocolate, candy corn, and hot chocolate mix. Candy provides the sugar that cows would usually get from corn, giving them more energy and making them become fatter. Lastly, chicken manure, ground limestone, and crab guts are also all used to fatten up cows while trying to keep feed costs down.[79]

TRUE COST ECONOMICS

True cost economics is an economic model that seeks to include the cost of negative externalities into the pricing of goods and services. An external cost occurs when producing or consuming a good or service then imposes a cost upon a third party. Products and services can directly or indirectly cause harmful consequences to living beings and/or the environment, and these must be paid for eventually. If one factors in these costs, the true cost of a product or service can then be more precisely determined. Examples of external costs may include the cost of disposing of the product at the end of its useful life; the environmental degradation caused by its emissions, pollutants and wastes from its production; or social costs associated with rising unemployment due to increased automation. Extreme externalizing of the full costs of a product or service is considered market failure because everyone will have to pay later for something that everyone did not consume today. In other words, market failure is a situation in which the allocation of goods and services by a free market is not efficient, often leading to a net loss for society at large.

Economists consider climate change to be a market failure, because it imposes huge costs and risks on future generations who will have to suffer the consequences of our actions. These costs and risks are not normally reflected in current market prices. To overcome this market failure, they argue, we need to internalize the costs of future environmental damage by putting a price on what causes that damage – namely, carbon emissions.

With this view in mind, let's look briefly at the economic cost of a burger in terms of greenhouse gas pollution. It is difficult to put a precise price on the cost of this carbon, but the U.S. Government's official monetary valuation (cost) of greenhouse gas pollution is roughly \$37 per metric ton of CO_2 emissions. Other sources put the figure ten times higher. Using the standard calculations of beef production discussed here in the section on the environment, the monetary cost of the carbon emissions produced by the average cheeseburger might range from 15 cents (the official U.S. government rate), to 24 cents (conservative independent sources) and even \$1.20 (high independent). Taking the average of these three estimates, however, would add 53 cents to the cost of a typical burger.[80]

Now let's look at the substantial direct (yet external) health-care costs of a mammal-centric diet and the indirect costs that result from unpaid care from family or friends and lost work days. There is an enormous economic burden placed on society from obesity and a handful of deadly chronic diseases — arthritis, cardiovascular disease, hypertension, Type 2 diabetes, and some cancers, among others — produced by a diet heavy in mammal meat. Direct medical diet-related costs are currently pegged at about $231 billion annually. Of that amount, by one estimate, the U.S. spent $190 billion on obesity-related health care expenses in 2005 alone.[81]

In a report published in the *Proceedings of the National Academy of Sciences*, Marco Springmann and his colleagues at the University of Oxford conducted an extensive study to determine the true cost that society must bear if it continues to center its diet on the consumption of mammals. Springmann and his team projected several possible dietary scenarios forward to 2050, comparing health-care and climate-related costs. They found that if the world continues its current meat-heavy diet, and does not shift to a diet that meets standard global dietary guidelines, the cost to the global economy could be up to $1.6 trillion a year by 2050. [82]

Of all the world's countries, the U.S. would save the most by further curbing its taste for meat. Due to its very high per capita health-care costs, it is estimated the U.S. could save $180 billion per year if its people ate according to recommended guidelines, and upwards of $250 billion if they avoided animal food products altogether—more than China or all of Europe combined. This calculation does not even include the number of obesity- and chronic-disease-related deaths that could be averted (at least 320,000 per year), and the accompanying benefits of reducing the level of greenhouse-gas emissions.

Springmann and his team made another estimate using a somewhat less intuitive measure called the "value of a statistical life," which put the savings from not eating meat in the neighborhood of $2 to $3 trillion in the U.S. and $20 to $30 trillion worldwide, per year.

Figure #1: The economic insanity of beef production

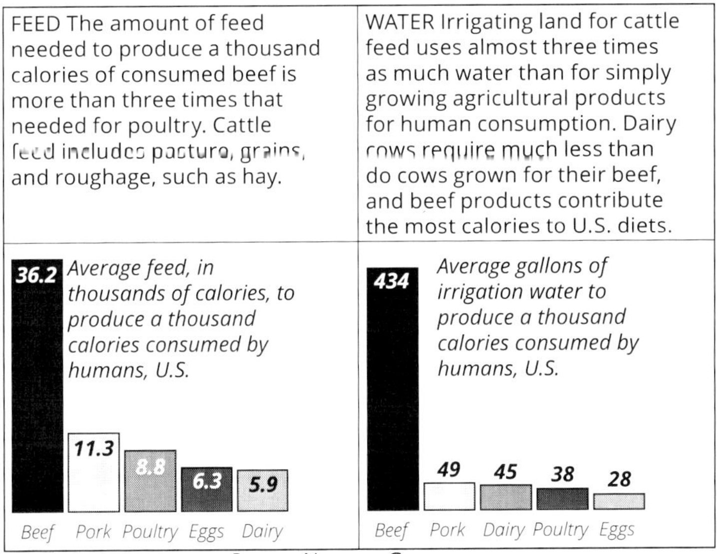

Source: National Geographic

THE COSTS OF CONTAMINATED WATER

There are other costs that are not so easy to calculate, but which are nevertheless real and should be included in any discussion of the true economic costs to society of its support of a mammal-meat centered diet.

Consider, for example, the elevated nitrates in water supplies that result from the chemical fertilizers used to grow the corn that feeds cattle. Nitrates now contaminate the public water supplies of almost 1,700 communities across the United States at levels the National Cancer Institute says could increase the risk of cancer.[83] Federal data shows that about two-thirds of those – 1,155 systems' serving more than three million people – have no treatment systems in place to reduce nitrate concentrations to safer levels.

This phenomenon creates costs to society that inevitably must be paid somewhere by somebody. Either households purchase bottled water, the costs get transferred to the healthcare sector when people fall sick, or local governments respond with higher taxes to clean up the mess.

Removing nitrates from tap water is expensive, and these costs must be absorbed by direct local taxation, municipal bonds, or the increased cost of water. The city of Des Moines recently had to spend $3.7 million to build a water treatment facility for precisely this reason.[84] In October of 2017, Hiawatha, Kansas, built a plant for $3.5 million to deal with nitrate levels that were so high that residents were warned not to drink the tap water.[85] In 2005, the City of Chino, California, spent $4.6 million on an ion exchange system to deal with its dangerously high nitrate levels.[86]

WASTED OPPORTUNITIES

According to figures drawn from the (British) Institution of Mechanical Engineers, as much as 50 percent of all food produced in the world ends up as waste every year.[87]

Such poor economic planning is particularly troublesome to me, considering the number of people who go to bed hungry every night. The UN Food and Agriculture Organization has estimated that of the 7.6 billion people in the world today, about 815 million suffer from chronic undernourishment and are thus unable to lead a healthy active life. The number of hungry people in the world is also growing – now reaching one in every nine people - while limited progress is being made in addressing the multiple forms of malnutrition.

One-third of the food produced in the world for human consumption every year — approximately 1.3 billion tons — is lost or wasted.

According to the UNFAO, roughly one- third of the food produced in the world for human consumption every year — approximately 1.3 billion tons — is lost or wasted. The average person in North America and Europe wastes up to 253 pounds (115 kg) of food a year, while consumers in sub-Saharan Africa, South and South-eastern Asia, each throw away 24 pounds a year.[88] Aside from the tragedy that is apparent in the face of starvation, this waste is also a major squandering of resources, including water, land, energy, labor and capital, and further still, it needlessly produces greenhouse gas emissions, which contribute to global warming and climate change.

As we have seen, nearly 60 percent of the world's agricultural land is now used for beef production, and yet beef accounts for less than two percent of the calories consumed worldwide. By producing all this beef, we are wasting an opportunity to feed those who do not have enough to eat. By feeding enormous amounts of agricultural products to cows, so we can in turn eat the meat on those cows, we are – knowingly or unknowingly – missing an opportunity to grow and provide food to those who go to bed hungry every night. Worse, we can see that by seeking to cut corners and thus feed dead cows to cows, we ended up having to kill and burn the carcasses of 4.4 million cows – for nothing. We can do better. We must do better.

REASON 4:

WE ARE ALSO MAMMALS

The ***FOURTH*** *reason to reduce or eliminate mammal consumption from your diet is because we share a special and close relationship with all mammals*

IN THIS FINAL CHAPTER, I will try and make the case that we should reduce or eliminate our consumption of mammals because we, as mammals, should look after one another. They've got our back, and we should have theirs – to protect them, not eat them. We mammals are more similar to each other and connected to each other than most of us realize – and to eat them may be just one step away from cannibalism.

You don't have to look far to read instances of cases where mammals take care of each other with very human characteristics, nor to find evidence of other mammals reaching out to save and protect humans. In this chapter I will first present some cases from the real world before taking a quick walk through the animated imaginations of our childhood, where we actually bonded quite closely with a wide variety of our mammal friends.

GENETIC SIMILARITIES AMONG MAMMALS

In the grand scheme of human history, our knowledge of evolution and DNA sequencing is only a relatively recent discovery. This science has enabled us to see how closely related we are as a species to every other living thing on this planet. Genetically speaking, we have much more in common with our fellow mammals than we do with the rest of life on Earth. Our closest genetic matches are of course our fellow humans with whom we share 99.9 percent of our genes.

The chimpanzee and another ape, the bonobo, are humans' closest living relatives.[89] We look alike in many ways, both in body and behavior – probably because we share a surprising 98.8 percent of our DNA with them.

Ongoing research is now discovering that we are also extremely close to the pig. A study funded by the European Research Council in 2011 discovered that the pig and its cousin, the wild boar, share more DNA similarity with us than we do with a mouse, a dog, a horse or a cow.[90]

Medical researchers were not surprised to discover our close kinship with pigs. We already are using parts of pigs for various medical conditions. For example, insulin is typically derived from pigs for injection into humans, and estrogen from pigs is used for birth control pills. Plastic surgeons also use dermal fillers made from pig fat. Severe burns are often treated with skin grafts derived from the epidermis of pigs. Researchers have also discovered that when human genes are combined with pig genes, they appear to fuse seamlessly..[91]

Dogs might be man's best friends, but they also happen to share 84 percent of our DNA.[92] Cat people will love hearing that their feline friends are even more like us than dogs are. A 2007 study found that about 90 percent of the genes in the Abyssinian domestic cat are similar to humans. Domesticated cattle share about 80 percent of their genes with humans, according to the Bovine Genome Sequencing Project, an international consortium of more than 300 scientists from 25 countries who worked six years to complete the genetic sequence of the beef cow in 2009. [93] In these instances, 80 percent seems to be the dividing line for us mammals, because when you drop much lower you get to chickens, at 65 percent.

HEAD HUGS

One of the first things I noticed about our fellow mammals is the way they – and we – greet each other. I have not seen this same rapport in other species of animals, only mammals, giving rise to the possibility that all mammals possess a consciousness much like our own and a deep awareness of the bonds we share with one another.

All mammals possess a consciousness much like our own and a deep awareness of the bonds we share with one another.

For the Maori people of the South Pacific, the traditional greeting between two people involves pressing noses and foreheads together. I like to call this a head hug – they call it a *hongi*. The greeting is used at traditional meetings among the Maori and at major ceremonies. In the *hongi*, the *ha* (breath of life) is exchanged in a symbolic show of unity. Through the exchange of this greeting, one is no longer considered *manuhiri*, a visitor, but rather *tangata whenua*, one of the people of the land.[94]

For the Inuit of the far north, including the Kalaallit in Greenland, the Inuvialuit in Canada, and the Inupiaq, Yuplit, and Alutiiq in Alaska, rubbing noses with a loved one is called a *kunik*. It is an expression of affection that involves pressing the nose and upper lip against the skin (commonly the forehead) and breathing in.[95] In the Socotra archipelago of Yemen, men touch their noses once or three times when greeting, as a gesture of both friendship and respect.

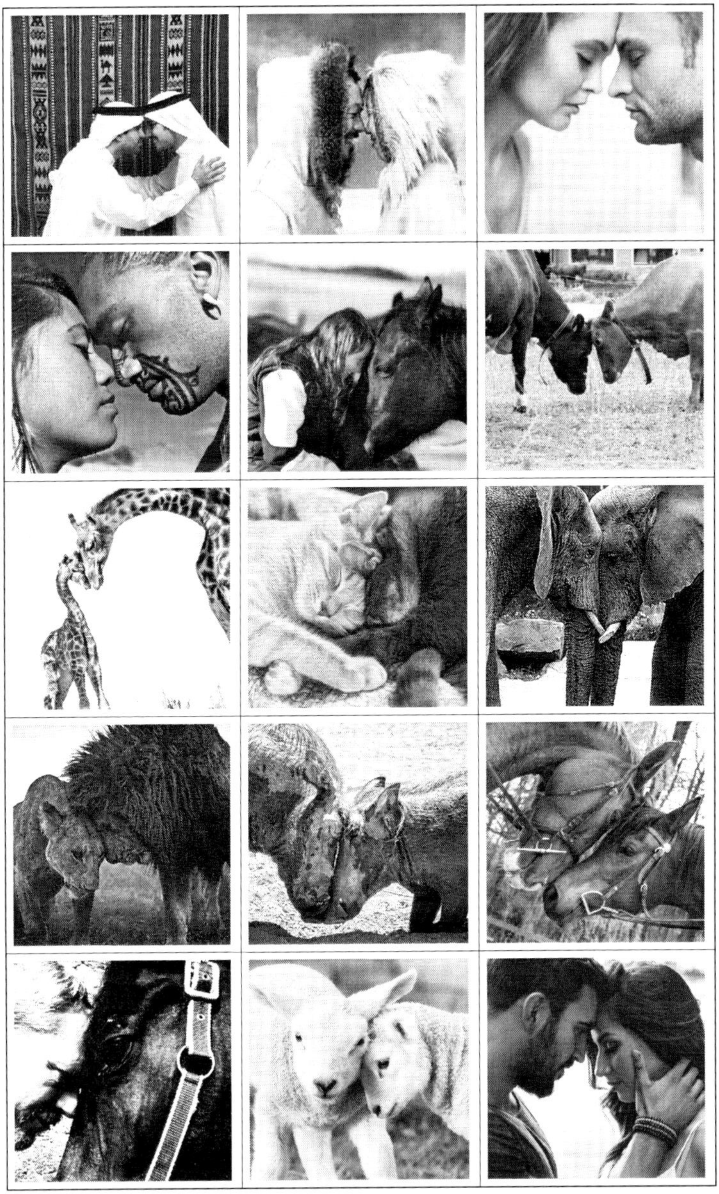

Mongolian nomads of the Gobi Desert have a similar practice, as do certain Southeast Asian cultures, such as the Bengalis, Cambodians, Laotians, Thai, Vietnamese, Timorese, the Sabu people, Sumbese and the Ibans. Arabs also press the bridges of their noses together as a traditional greeting when meeting members of the same tribe, family, or clan. Many Buddhists and Hindus believe we have a third eye in the middle of our foreheads and connecting these eyes with others opens a channel for an exchange of energy and unspoken communication.

These greeting rituals also seem to be common among other mammals, from cats and dogs to giraffes and elephants. I have seen mammals engage in head hugs throughout the world -- giraffes, cows, and moose – even elephants.

If you have ever seen a female dog approach a litter of pups that are not her own, you may notice she lowers her head and touches the nose of each with her nose. She nuzzles the faces of some a bit and sniffs other parts of their bodies; however, the opening contact is almost always a nose-to-nose touch. Psychologists who study animal communication say this snout contact appears to be part of a greeting ritual.[96] Cats use this nose touch greeting virtually with any cat they meet that appears to be non-threatening.

If you have not yet experienced a head hug with another mammal – try it (both with a significant other human and with any of the animals pictured below), you may be surprised at the result.

MAMMALS TO THE RESCUE

For much of my life, I have seen stories in the media about mammals rescuing humans. Each time I made a mental note, thinking, "hmmm that's interesting," but I never really focused on it until these stories started to mount up over time. While some of the mammals in this section are obviously not in danger of being eaten, my point here is to establish a new paradigm in which all mammals are part of a family, *our* family, and thus, should be included within our circles of compassion.

While traveling through Bangladesh in the 1980s, I read a story in the local newspaper about a boy who had been taken far out to sea in a hurricane when a tsunami washed in over his village. Two days later, the boy was spotted by a fisherman -- balanced on the backs of two dolphins. I tried to find that story, but I couldn't. However, in searching for it I did come across many others.

"THAT'LL DO PIG, THAT'LL DO"

A study at Emory University in the United Kingdom revealed that pigs share several cognitive capacities with other highly intelligent species, including dogs, chimpanzees, elephants, dolphins, and humans. This research suggests that pigs have excellent long-term memories and are skilled at mazes and other tests that require the location of objects. Moreover, they can build relationships with each other and also other fellow mammals.[97]

Pigs have excellent long-term memories and are skilled at mazes and other tests.

In the Midwestern United States in 1997, Jo Ann and Jack Altsman agreed to babysit their daughter's Vietnamese pot-bellied pig, Lulu, while the daughter was on vacation. They grew fond of the pig and decided to keep it at their home in Pennsylvania. Lulu grew and grew, from four pounds to 150 pounds a year later. In August 1998, Jack was fishing on Lake Erie when Jo Ann, 61, had a heart attack-- her second in eighteen months. She threw an alarm clock through the window hoping to attract attention, but to no avail. Their dog, Bear, an American Eskimo, just barked.

According to the news reports, [98] at first Lulu "cried big fat tears" but then she took action, squeezing her obese body through the doggy door and pushing open the gate. Whenever a car came, she would lie in the road in front of the approaching vehicle. This activity went on for about 45 minutes as car after car simply drove around her. Finally, a car stopped, and the driver followed LuLu into the house and saw Jo Anne unconscious on the floor. The motorist immediately called 911, and a helicopter was dispatched. Jo Ann was flown to the hospital where the doctors were quoted as saying that had she arrived fifteen minutes later, she would have died.

DOGS AS MAN'S BEST FRIEND

The audience for this book probably does not need much convincing to keep dogs off the dinner plate. However, dogs are mammals, and as such, are genetically close to all the mammals on the menu at the Black Angus Steakhouse, and thus worthy of a mention here in our discussion about bonding with mammals.

The reason we call dogs "man's best friend" needs little explanation, at least in Western countries. As humans, we take great pride in our dogs. We often show them off to others just as we do our own children. We play with our dogs, hold them close, and rely on them for many reasons, including security, friendship, and compassion. We rely on dogs also for law enforcement and military operations.

Every dog owner knows there are benefits to having a dog, from barking at intruders to loyal companionship. However, for people with disabilities, the presence of a dog can be critical to their daily functioning. The emotional support and comfort provided by their pets allows them to deal with challenges that might otherwise compromise their overall quality of life. These pets are known as service and emotional support animals. They also can be trained to be psychiatric service dogs to support people whose disability comes from a mental illness. These dogs can detect the beginning of psychiatric episodes and help ease their effects. There is little wonder why Brigitte Bardot dedicated her life to protecting them.

"BEARLY" SURVIVED

In January 2019, three-year-old Casey Hathaway wandered away from his grandmother's backyard in North Carolina and became lost in the woods. A local emergency team was called that involved helicopters, drones, K-9 units, and divers, as well as hundreds of volunteers to search for the toddler, who was not dressed for the brutal below-freezing weather conditions. The conditions were indeed so bad that the search had to be called off after nightfall.

Two days later rescuers found him tangled up in thorny bushes, cold and soaked, but safe. When asked how he had survived, the little boy said a bear had come and kept him warm and safe.[99] North Carolina is home to many black bears. The adults did not believe him at first, but the little man stuck to his story.

DOLPHINS – MAN'S OCEANIC PROTECTOR

Dolphins are known to be some of the most compassionate and intelligent animals of all mammals. There is recorded evidence of them helping humans throughout history. In ancient Greece, many coins show images of children, men, or gods riding a dolphin. According to legend, the sea god, Poseidon, sent a dolphin to save his son, Taras, from a shipwreck.

When the surfer Todd Endris was attacked by a 15-foot white shark off the coast of California in the summer of 2007, dolphins came to his rescue. The shark had taken a bite from his back and sunk its teeth into Todd's leg when a pod of dolphins suddenly appeared and attacked the shark, then forming a protective ring around him until the shark left. Todd was then able to catch a wave back to shore and seek emergency treatment[100].

A pod of dolphins suddenly appeared and attacked the shark.

The American actor,Dick Van Dyke, was once saved by porpoises after he drifted out into the Atlantic Ocean on his surfboard. *Time Magazine*[101] reported at the time that the actor had drifted so far out he could no longer see the shore. Van Dyke began a feeble attempt to paddle back in the direction from which he believed he had drifted, when he saw several fins protruding from the water that started circling around him. He first thought they were sharks, but then was relieved to discover he was wrong. "They turned out to be porpoises, and they pushed me all the way to shore," he said.

In 1996, a British tourist, Martin Richardson, was reportedly attacked by a shark in Egypt while swimming off the coast of Sinai in the Gulf of Aqaba. A group of dolphins surrounded him and kept the shark at bay by slapping the surface of the water until Martin could be pulled to safety.[102]

The British lifeguard, Rob Howes and his daughters were swimming 100 yards out to sea at Ocean Beach, near Whangarei, New Zealand when seven bottlenose dolphins suddenly sped toward them and herded the family together.

Howes tried to swim away from the group, but two of the bigger dolphins pushed him back. He realized what was happening only when he spotted a 10-foot great white shark heading straight toward them. The dolphins continued their defensive maneuvers for 40 minutes until the group could swim safely back to shore.[103]

Ingrid Visser, of Orca Research, an environmental group, was quoted in *The Telegraph*[104] newspaper in Britain as saying the dolphins' behavior was understandable, as they attack sharks to protect themselves and their young. "They could have sensed the danger to the swimmers and taken action to protect them," she explained.

There are many more instances of dolphins rescuing humans, but before this book turns into a treatise about dolphins, since they are not usually found on the menu, let's move on to other mammals.

GORILLAS IN THE MIST

In August 1996, a crowd of visitors at the Brookfield Zoo in Illinois, looked on in horror as they saw a three-year old toddler tumble more than 15 feet into the enclosure there and land near seven gorillas. Those witnessing the incident expected the worst. However, much to their surprise, an unlikely hero emerged. According to *The Chicago Tribune*,[105] Binti-Jua, a rare Western lowland gorilla, lumbered over to the boy and protected him from the other apes. She cradled him in her arms, carried him to a doorway, and laid him gingerly at the feet of the waiting paramedics.

This case isn't an isolated one. Ten years earlier at the Jersey Zoo in the United Kingdom, a five-year-old boy fell into a gorilla enclosure and lost consciousness. A large male gorilla named Jambo stood guard over him, not allowing any of the others to come near. When the boy woke up and started crying, all the gorillas backed off, and zookeepers (along with an ambulance) were able to retrieve him safely.[106]

LIONHEARTED PROTECTORS

In rural southwest Ethiopia in June 2005, a 12-year-old girl was snatched by four men on her way home from school. Such abductions, which often lead to forced marriage, are still common in those rural areas. However, this time, it ended differently. A week after she was taken, the local police managed to track her down and were in hot pursuit when the kidnappers fled – taking the girl with them. During the escape, three lions encountered the group and chased the men off. The lions stayed with the girl without harming her, before they departed when police searching for her came near. Sergeant Wondmu Wedaj told the BBC[107] they found the girl alive but shocked and terrified. An Ethiopian wildlife expert was quoted in the same article as saying the lions could have spared the girl "because her crying may have sounded like the mewing sound from a lion cub." The four men were later caught by the police and tried for kidnapping.

NOT HORSING AROUND

It was a beautiful summer day in Castle Douglas, England, toward the end of July 2014, but for Fiona Boyd it was just another day on the farm. She had to move a cow and its calf away from the paddock, as there were other cows about to give birth there.

Farmers normally separate dairy cows from their calves when they are very young. After birth, they walk both mother and calf down to a special barn, and then afterwards, the mother is returned, on her own, to the milking parlor, without her newborn.

On this day, Fiona began to walk behind the two-day-old calf and cross over the field. "It was a little thing with its back reaching my knees," she wrote in an editorial published in the British newspaper, *The Guardian.*[108] "Normally we'd expect the mother to just happily follow."

Not this time. The calf started to get a bit agitated. Mom was upset and, in a panic, ploughed into Fiona's left side, knocking her to the ground. The frantic woman screamed, but there was no one at the farm to help her. She knew that once someone is on the ground, dairy cows have been known to group together and attack en masse – and kill. She knew she had to get back on her feet.

However, the cow wouldn't let her get back up. She kept pushing Fiona back onto the ground with her head. "I shouted and tried to hit the beast's face," Fiona said, "but she was so powerful I couldn't push her away. She was straddling me, and all I could see was her massive body looming over me. Instinctively, I knew she was about to crush me."

Fortunately, Fiona's horse, Kerry, was nearby and galloped to the rescue, lashing out at the cow, while Fiona crawled away to safety under an electric fence. Needless to say, Kerry was Fiona's new best friend and was pampered until her death from natural causes two years later.

"HOW NOW, BROWN COW?"

Cows exhibit a wide array of personalities and characteristics, similar to dogs, cats, and even people. They are generally quite intelligent animals who can remember things for a very long time. Animal behaviorists have found that they interact in socially complex ways, develop friendships over time and sometimes even hold grudges against other cows that treat them badly.[109]

Cows exhibit a wide array of personalities and characteristics.

These gentle giants mourn the deaths of and even any separation from those they love, sometimes shedding tears over their loss. The mother/calf bond is particularly strong, and there are countless reports of mother cows who continue to call and search frantically for their babies even after the calves have been taken away and sold to veal or beef farms.

Janice Wolf is the founder of Rocky Ridge Refuge, a sanctuary in Arkansas that cares for animals with special needs. One day she was in the back pasture of the refuge when an 11-month old Ankole-Watusi calf (a modern American breed of domestic cattle) suddenly turned and blocked her path. She couldn't understand why, so she took hold of its horns and tried pushing it away, but the calf tossed its head and knocked her off balance. That is when Janice spotted a copper-head snake on the ground exactly where her foot would have gone had the baby cow not intervened.[110]

In New Zealand, a farmer's wife who was swept away by floods had her life saved by a cow. Kim Riley praised the animal - known only as Number 569 - and described it affectionately as "an ugly old tart". She was leading a herd of 350 cows in early morning darkness when they were caught in a torrent of floodwaters. The cows who could do so swam to an outcrop of dry land, abandoning Riley to her fate. One, however, turned back to get her. "I looked back and saw one of the last cows bearing down on me, number 569," she later recalled. "I threw my arm over her neck... the warmth that was coming from her was so reassuring. I just laid back and relaxed, and said, 'Take me home.'"[111]

RABBIT TO THE RESCUE

Simon Steggall of Warboys, Cambridgeshire, United Kingdom, had been a diabetic since childhood. He had to inject himself with insulin four times every day. One day in 2004, Simon slipped into a coma while watching television. His wife, Victoria, was in the next room and thought he had simply nodded off. His pet Dory, a large rabbit weighing 21 pounds, immediately sensed something was wrong. When Simon slipped into a coma, Dory jumped up on his chest and thumped furiously on his face. His wife, who was a nurse, noticed something was wrong, and ran to his side, attempting unsuccessfully to revive him. She then called the paramedics, who rushed him to the hospital, where he received the medical care he needed to save his life. The doctors said that if he had stayed in the coma for much longer he would have died.[112]

THE MEMORY OF AN ELEPHANT

Elephants help each other in distress, grieve for their dead, and share emotions with one another—just like we do. Asian elephants show concern for others when they are stressed by using their trunks to caress the suffering elephant. Scientists have spotted elephants assisting others that are injured, plucking out tranquilizing darts from their companions, and spraying dust on the wounds of their friends. While doing research for this book I came across a report of an elephant who struggled to help a dying friend, lifting her with her tusks and trunk, while calling out in distress.[113]

Asian elephants, like great apes and dogs, recognize when a herd mate is upset. They have been documented offering gentle caresses and chirps of sympathy. Consoling elephants often put their trunk into another elephant's mouth to communicate vulnerability, perhaps showing the other elephant that they are there to help.[114]

Elephants, like most mammals, are highly altruistic animals. They will aid other species, including humans, who are in distress. Once in India, an elephant was helping locals lift logs by following a truck and placing the logs in pre-dug holes upon instruction from the *mahout* (an elephant trainer). At one certain hole, the elephant refused to lower the log. The *mahout* came to investigate the hold-up and noticed a dog sleeping in the hole. The elephant would only lower the log when the dog was gone.[115]

In April 2019, a group of poachers illegally entered Kruger National Park in search of rhinoceros. Rhino poaching has escalated in recent years and is being driven by the demand for rhino horn in Asian countries, particularly Viet Nam. It is used in traditional Chinese medicine, but increasingly common is its use as a status symbol to display success and wealth. South Africa holds nearly 80 percent of the world's rhinos and has been the country hit hardest by poaching criminals, with more than 1,000 rhinos killed each year between 2013 and 2017.

On this particular day, however, as the poachers tried to chase down a rhino, they ran into some bad luck in the form of an elephant. The elephant attacked the group, knocking one of them to the ground. The others ran away as their partner-in-crime was trampled to death. To finish it off, more mammals – lions – joined the party and ate the poacher. The accomplices informed the man's family of their loss. They then told the authorities, who went off in search of the man. A search party struggled to find the body but eventually found part of a human skull and a pair of trousers.[116]

OUR FASCINATION WITH ANIMATED MAMMALS

In logic, *reductio ad absurdum* (Latin for "reduction to absurdity"), also known as *argumentum ad absurdum* ("argument to absurdity") is a type of argument that attempts either to disprove a statement by showing it inevitably leads to a ridiculous, absurd, or impractical conclusion or to prove the statement by showing that if it were not true, the result would be absurd or impossible.

The two mammals that immediately come to mind as being both food and friend are Porky Pig and Clarabelle the cow.

If you think about it, the fact that we can love and cherish the same animals we also enslave, torture, kill and consume is the ultimate *reductio ad absurdum.* Mammals have been our best friends since childhood – most cartoon characters depicting animals are in fact mammals – although the mammals we choose to eat are often less famous. They have, nevertheless, lit the imaginations of children for generations, making one wonder how we can grow up and eat them.

The two mammals that immediately come to mind as being both food and friend are Porky Pig and Clarabelle the cow. There also is the operatic cow, Ermintrude, and of course Gladys, who is the most famous of Sesame Street's many bovines. Peppa Pig is a small British pig with a red dress and a very loud "oink." She lives with her baby brother George, Mummy Pig and the unshaven and aptly-named Daddy Pig. Other mammals of repute that appear around religious occasions are of course the Easter Bunny and Rudolph the Red-Nosed Reindeer.

Piglet is Winnie the Pooh's closest friend. Although he is a "very small animal" with a generally timid disposition, he tries to be brave and on occasion even conquers his fears. Pooh is an anthropomorphic, soft-voiced, cuddly, loveable, and quiet toy bear in the company of a group of like-minded mammals who fill Christopher Robin's imagination and provide sage advice and perspicacious perspectives. There is Eeyore, Pooh's ever-glum, sarcastic, and pessimistic donkey friend. Kanga is a female kangaroo and the doting mother of Roo. Rabbit and Tigger round out the crew, except for the one non-mammal, the aptly named Owl.

Possibly the only thing better than a great story is a great story with memorable mammals. There are many unforgettable literary animal characters that have made lasting impressions on us and taught us all how to be better humans through their examples of bravery and loyalty. If you think about it, many of our childhood friends were mammals, cartoon characters, pets, and the stuffed animals we hugged when we fell asleep.

Sure, there is also Dory the fish, Harry Potter's pet snowy owl, Hedwig, the adorable Rango, the chameleon -- and who can forget Leonardo, Donatello, Michelangelo, and Raphael of the Teenage Mutant Ninja Turtles? And yes, there is the spider Charlotte, who saves her mammal companion (a pig) from the dinner table by spinning messages in her web. There we find a definite message that mammals are better kept as friends than served as dinner.

But while Charlotte spins her web, Aslan, the ruler of *Narnia*, is a majestic lion who is the embodiment of goodness and justice. Simba, in *The Lion King*, grows up to take his father's place as the King of the Pride Lands. Rafiki (meaning "friend "in Swahili) is his friend and another mammal, who lives in a *baobab* tree in the Pride Lands and performs shamanistic rituals for the lions of Pride Rock.

Bagheera guides and protects Mowgli from the other animals in the jungle – many of whom like Shere Khan are mammals while others, like Kaa, the massive python who remains intent on hypnotizing a captured Mowgli, are not. Baloo is a sleepy old brown bear, who teaches Mowgli the law of the jungle. Mammals helping mammals is always the theme.

Nearly all our favorite characters in *Madagascar* are mammals. Alex is a male African lion who is best friends with Marty the zebra, and also friends with Gloria, the hippopotamus, and Melman, the giraffe. After escaping from the zoo and washing up on the beaches of Africa, Alex becomes extremely hungry after days of being unable to eat. Hunger drives Alex into a crazed state, almost killing Marty in hunger, not realizing that steak is animal flesh. Alex is given sushi by the penguins to satiate his hunger. Mammals, who are given fish to eat by birds.

There is also Bugs Bunny, a mammal with mostly mammal friends like Porky Pig, Elmer Fudd, Sylvester and Wile E. Coyote (and also there are a few non-mammals like Daffy Duck, Tweety and of course the obnoxious Road Runner).

TH-TH-TH-THAT'S ALL FOLKS!

At the end of all Warner Bros' classic Looney Tunes cartoons, Porky the Pig appears at the closing screen proclaiming " Th-th-th-that's all folks!" The voice is that of Mel Blanc, the most prolific voice artist in the history of Hollywood. Blanc created voices for an estimated 90 percent of Warner characters, including such cartoon stars as Bugs Bunny, Daffy Duck, Porky Pig, Tweety Bird, Sylvester the Cat, Yosemite Sam, Foghorn Leghorn, Marvin the Martian, Pepé Le Pew, Speedy Gonzales, Wile E. Coyote, Road Runner, the Tasmanian Devil, and many of the other characters from the Looney Tunes and Merrie Melodies theatrical cartoons during the golden age of American animation. He voiced all of the major cartoon characters except for Elmer Fudd, whose voice was provided by fellow radio personality Arthur Q. Bryan. When Bryan died in 1959, Blanc took over voicing Fudd, as well, as there was no one around with enough talent to do it.

Blanc was incredibly strong and healthy for much of his life, despite the fact that he started smoking as a teenager and smoked a pack of cigarettes a day. In 1961 he survived a near-fatal car accident that put him in a coma for a few weeks, breaking many of his bones. At 77 he suffered from emphysema, but recovered remarkably after giving up smoking. In the end it was advanced coronary artery disease that took his life at the age of 81.

REASON 5:

GLOBAL COORDINATED EFFORTS TO ADDRESS THE CHALLENGES

EVERY COUNTRY IN THE WORLD is currently engaged in a coordinated response to the environmental, social and economic challenges that are plaguing our planet. Whether or not we succeed of course remains to be seen, and depends– in no small measure – on you.

After several years of intergovernmental deliberations and community consultations among civil society groups the United Nations produced the 17 Sustainable Development Goals (SDGs) in 2015. The SDGs contain 169 quantified targets and their progress is measured by 100 specific indicators. All have a deadline of 2030 and commit all governments to "comprehensive, integrated, and universal transformations." Countries are expected to mobilize their full efforts to end all forms of poverty everywhere, fight inequalities, and tackle climate change, while also ensuring that "no one is left behind."

The SDGs are in part an extension of the successful eight Millennium Development Goals (MDGs) that were slightly less ambitious but nevertheless comprehensive (formulated at the turn of the millennium to be achieved by the year 2015). The first of the 8 MDGs, for example, was to reduce extreme poverty by half – a goal that was achieved ahead of schedule.

This time, by adopting the 2030 Agenda for Sustainable Development (Also known as "Agenda 2030" and in which the SDGs are embedded), the international community has committed itself to completely eradicating hunger and poverty and achieving other vital goals, such as making agriculture sustainable, securing healthy lives and full employment with "decent work," reducing inequality, and making economic growth more inclusive so that all can benefit.

While much of the responsibility lies with governments to enact policies and enforce regulations, we will never make it unless you, the consumer, becomes more aware of the consequences and repercussions of your choices – especially those of your dietary decisions.

FOOD AND THE SDGS

Both food and agriculture feature prominently in the SDGs (also known as the "Global Goals"), because the two are interconnected and involve almost all aspects of the economy, the environment and society:

freedom from hunger, malnutrition, desertification, sustainable water use, loss of biodiversity, and also overconsumption, obesity and public health. On a personal level, by consciously looking at the way we eat, we can responsibly respond to all 17 goals and make a direct and personal contribution toward making this world a better place for all.

We can make huge strides toward making the world healthier, wealthier and happier simply by reducing or eliminating mammal consumption.

As we have seen here already, what we choose to eat, as well as our production and cultivation processes that produce our food, are having a devastating impact on our planet and the people who live on it. We can make huge strides toward making the world healthier, wealthier and happier simply by reducing or eliminating mammal consumption – especially those mammals who have been produced *en masse* using unsustainable industrial processes.

A powerful report from the panel of nutrition, agriculture, and environmental experts at EAT, a Stockholm-based nonprofit that seeks to improve the world food system, was published in January 2019 in the medical journal *The Lancet.* The report declared that "getting it right with food will be an important way for countries to achieve the targets of the UN Sustainable Development Goals and the Paris Agreement regarding climate change."[117]

The report, *Food in the Anthropocene: The EAT-Lancet Commission on healthy diets from sustainable food systems*, declared that if we are to survive as a species, everyone MUST choose to eat mostly vegetables, grains, legumes, and nuts. For those who choose to eat mammal meat, the report says it would be better for them to limit it to just one serving per week.

A "planetary health diet," as the authors call it, was a provocative recommendation, especially for those countries where many people eat multiple servings of meat each day. Achieving the vision of this report will require a radical revamping of our food culture — indeed prioritizing sustainability and collective survival over entrenched traditions and habit.

The following is a quick look at the 17 SDGs, through the lens of how they relate to the vision outlined here in *No Mammal Manifesto*. For more detailed information on these Global Goals, search for them on Google – there is a growing mountain of information from groups who are actively committed to their achievement. If you are on twitter, follow the hashtag #SDGs.

GOAL 1: NO POVERTY

More than 700 million people, or 11 percent of the world population, still live in extreme poverty and are struggling to fulfil the most basic needs like health, education, and access to water and sanitation. Eradicating poverty in all its forms remains one of the greatest challenges that has faced humanity. While the number of people living in extreme poverty did drop by more than half between 1990 and 2015 – from 1.9 billion to 836 million (and have dropped still further to 815 million as of 2019) – too many are still struggling for their most basic human needs.[118] Proper government policies that promote public investment in food security will enable and then reinforce this key goal, since enhanced food and nutrition security are essential to reduce poverty and indeed eradicate extreme poverty. For all these reasons, it is clear that any government policies that allow for inefficient caloric conversions, food waste, and the destruction of agricultural land should be avoided and replaced with pro-poor policies that work to ensure no one is left behind.

Eradicating poverty in all its forms remains one of the greatest challenges that has faced humanity.

GOAL 2: ZERO HUNGER

Today, 815 million people are hungry, and every third person is malnourished, clearly reflecting a food system out of balance.[119] The SDGs aim to end all forms of hunger and malnutrition by 2030 by making sure all people – especially children – have ongoing access to sufficient and nutritious food year round. The Agenda recognizes that ending hunger will require "sustainable agricultural practices." It highlights that these efforts in turn, will necessitate the support of small-scale farmers and allow equal access to land, technology, and markets.

To meet the world's future food security and sustainability needs, food production must grow substantially while, at the same time, agriculture's environmental footprint must shrink significantly. Achieving the goal of ending hunger will require a complete rethinking of how our food systems work. For example, tremendous progress can be made by halting agricultural expansion, closing 'yield gaps' on underperforming lands, increasing crop efficiency, shifting diets and reducing waste. Together, these strategies could double food production while greatly reducing the environmental impacts of agriculture occurring today.

Whatever approach we take, it is clear we need to produce enough healthy food and we need to do it sustainably, so that production remains secure well into the future.

For all the reasons outlined here in this Manifesto, promoting industrialized mammal meat production clearly is not a sustainable activity – by any stretch of the imagination. Thus, ending government subsidies for this sector will be important. Bolstering the agricultural sector to ensure food security is not in and of itself a bad idea, but any government support (as direct or indirect subsidies, tax breaks, market access etc.) should only be directed at activities that are sustainable. Clearly, raising cows, pigs, sheep, etc. is not sustainable.

GOAL 3: GOOD HEALTH AND WELL-BEING

More than six million children still die before their fifth birthday each year, and only half of all women in developing regions have access to the health care they need. While good health as a sustainable goal is primarily concerned with reducing infant mortality and providing proper healthcare where and when needed, it also addresses the need to keep all people healthy up to and through, adulthood. It recognizes that noncommunicable diseases are the biggest cause of premature death in the world today. Obesity and malnutrition are major culprits behind this epidemic of poor health, so switching to healthier diets at any age can turn things around, giving people longer, more enjoyable lives.

More than six million children still die before their fifth birthday each year.

As we have already discussed in the preceding pages, good health and well-being can be achieved by eliminating or reducing our consumption of beef, pork, mutton, veal and other mammals. It is not just eating mammal meat that is unhealthy – the industrial production of it is polluting our water, our air, and our bodies.

The first step in promoting a healthier lifestyle is to pay attention to what we are feeding ourselves, our families, and our communities.

GOAL 4: QUALITY EDUCATION

This fourth goal primarily concerns enrolling all kids into school and providing them with a good quality education. The world is on track to achieving this goal, as the total enrollment rate worldwide reached 91 percent in 2015.

Providing a good quality education also necessitates ensuring optimal nutrition. Poor nutrition affects school achievement and attendance, impacting learning at any age. At the same time, schools are an ideal setting to teach and provide students with more opportunities to practice healthy behaviors.

Undernutrition in children is linked to poor mental development and behavioral abnormalities.[120] In poor households, children may not be getting the nutrition they need, so schools are ideal places to give children at least one good meal a day. The education system is also the perfect place to raise more awareness about healthy as well as sustainable diets.

Thus, any effort to achieve Goal 4 should focus on teaching nutrition, while also ensuring students are eating well. As we have seen, eating well means avoiding foods that contain mammal meat. To set a good example and deliver the best possible nutrition to students, many governments offer assistance to schools to make sure their students are eating well. The National

School Lunch Program in the United States, for example, is a Federally-assisted meal program that operates in both public and nonprofit private schools and also residential child care institutions. It seeks to ensure that all school children in the United States receive a nutritious lunch every school day. The vast majority of schools — approximately 95 percent — participate in this program, thus providing meals to more than 30 million children on any given day.

Unfortunately, although a great idea in principle, there is still room for improvement. Students who eat a National School Lunch Program lunch in the United States are more likely to be obese compared to students who bring their lunch from home. Despite the ongoing warnings from many of the sources in this book, National School Lunch Program often serves processed beef and pork – often of such a poor quality that it has been turned down by fast food restaurants like McDonald's and Burger King for failing to meet their quality or safety standards.[121]

GOAL 5: GENDER EQUALITY

Women and girls represent half of the world's population and most probably more than half of its potential. Holding half of humanity back holds us all back. Unfortunately, gender inequality persists everywhere, stagnating social progress. Women and girls are more severely affected by poverty, hunger and disease than their male counterparts. When food is scarce, female family members often get the smallest portions. In terms of work, especially agricultural work, women (more than men) are often paid starvation wages or fall into indentured servitude – a modern form of slavery.

Holding half of humanity back holds us all back.

Agenda 2030 recognizes that ending all forms of discrimination against women and girls is not only a basic human right; it is also fundamental to accelerating and succeeding at sustainable development. Smart governments are starting to realize that empowering women and girls even has a significant multiplier effect, accelerating economic growth and development for everyone. A focus on ensuring that women and girls are eating a balanced and nutritional meal is a major part of this effort.

In the large developing countries such as India, China, and across Latin America, large-scale industrial meat production is displacing many small, rural producers. As an estimated 80 percent of growth in the livestock sector now comes from industrial production systems, livestock are in direct competition for scarce land, water and other natural resources. This situation is making it difficult for the rural poor, and when the going gets tough, women and girls tend to suffer the most. Thus, any efforts to make our food and agricultural systems more sustainable, practical, and logical will have a direct positive effect on expanding gender equality. This will necessarily involve a drastic reduction in the amount of land taken for industrial livestock production. If people are eating fewer hamburgers and steaks, demand for these products will decline, and the land grabbing should stop.

GOAL 6: CLEAN WATER AND SANITATION

Water scarcity affects more than 40 percent of people in the world – an alarming figure that is projected to increase with the rise of global temperatures from climate change. In 2011, according to the UN Development Program (UNDP), 41 countries experienced water stress. Of these, 10 are dangerously close to depleting their supply of renewable fresh water and now must rely on alternative sources. By 2050, it is projected that at least one in four people will be affected by recurring water shortages.

Ensuring universal access to safe and affordable drinking water for all by 2030 will require significant investments in adequate infrastructure, as well as the provision of more and better sanitation facilities. It will also necessitate reducing the industrial production and consumption of mammal meat.

We know that about 70 percent of freshwater is used for agriculture – and most of it is used to grow crops that are then fed to livestock. This form of intensive agriculture pollutes freshwater resources, especially through nutrient runoff. Sustainably managing water that goes into food production can speed up the goal of ensuring water availability.

GOAL 7: AFFORDABLE CLEAN ENERGY

One in seven people today have no light to turn on at night. Furthermore, more than 40 percent of the world's population, approximately three billion people, rely on polluting and unhealthy fuels for cooking. As energy is the dominant contributor to climate change, accounting for around 60 percent of global greenhouse gas emissions, any paths forward to bring energy to all must be renewable. Improving energy efficiency to avoid wasting energy is one solution. Studies show that adopting stronger efficiency standards could reduce global electricity consumption by buildings and industry by 14 percent.[122]

One in seven people today have no light to turn on at night.

Producing mammals for food is notoriously energy intensive. Reducing the amount of energy used in developed countries for food systems would be an important step in the right direction, as well as helping producers in poor countries source clean, reliable energy that they can use to increase more sustainable food production, closing the so-called yield gap and ensuring greater food security and improved nutrition overall for everyone everywhere.

As noted above, a massive percentage of agricultural land is used to feed cattle and hogs. If more of this land were used to feed chicken and fish – or better yet, feed humans directly, more crops could possibly be used for biofuels.

Energy crops include wheat, corn, soybeans and sugarcane. Biofuels burn cleaner than fossil fuels, releasing fewer pollutants and greenhouse gases like carbon dioxide into the atmosphere. While they are by no means a panacea for the energy crisis, biofuels are definitely sustainable, affordable, and clean. However, they should never be produced at the cost of food security, and their use should be carefully calibrated to meet demand price pressures on the cost of staple crops. My hope is that eventually new technology will be developed (or released) that generates infinitely clean and affordable energy from renewable resources. However, for the time being, we must do the best we can with what we have – and reforming our food production systems by eliminating mammal food sources will free up a huge amount of agricultural resources that can produce more biofuel.

GOAL 8: DECENT WORK AND ECONOMIC GROWTH

The SDGs promote sustained economic growth, higher levels of productivity, and technological innovation as a way of ensuring that everyone everywhere can find a good job – a job that lets them take care of their families while contributing to the overall well-being of society at large. Encouraging entrepreneurship and job creation are keys to this goal, as are effective measures to eradicate forced labor, slavery, and human trafficking.

Some of the world's most dangerous, grueling, and unhealthy jobs are currently in the food production and distribution systems. More sustainable food production can, and indeed must, ensure safer working conditions and better livelihoods for all food producers. The UN Food and Agriculture Organization estimates that more than a billion people work in agriculture alone, so positive change in the food sector will make a massive contribution to ensuring decent work for all.

Conversely, the development of new and more nutritious menus will create many opportunities, both in small-scale farming and also in the production and distribution of healthier alternatives to the status quo. The world is long overdue for disruption in this economic sector.

GOAL 9: INDUSTRY, INNOVATION, AND INFRASTRUCTURE

There are currently few good estimates of the total contributions that food, including production, processing, retail, and support functions, make to the global economy. However, World Bank data shows that the average consumer spends 40 percent of his/her income on food and beverages, and 30 percent of the labor force is employed in agriculture.[123] We know food is a huge and growing market. We've seen the power of food to drive innovation, including the coming "clean meat" revolution to try and produce meat through artificial means (covered in the next chapter). We also know that solving challenges related to infrastructure are an important component of building a more sustainable food system. Much of the food waste that occurs each year could be avoided through improved transport, processing, and storage after harvest.

UNFAO Director-General Jose Graziano de Silva has called for an "inclusive rural transformation" to leverage the enormous untapped potential of food systems to drive agro-industrial development, boost small-scale farmer productivity and incomes, and create off-farm employment in new and expanding segments. With rapid urbanization currently taking place in developing countries, we need to shift the dietary focus from mammal meat to healthier and more sustainable alternatives. I do believe this change will go a long way toward ensuring that truly inclusive transformation takes place, and more importantly, takes a firm and permanent hold as countries transition to higher levels of development.

GOAL 10: REDUCED INEQUALITIES

Income inequality is on the rise, with the richest 10 percent of the world's population earning up to 40 percent of total global income. The poorest 10 percent earn only about two percent of total global income. Reducing these ever-widening disparities will require the adoption of sound policies to empower the bottom percentile of income earners and promote economic inclusion of all people, no matter where they live.

To reduce inequalities between countries and within different societies, those at the bottom of the ladder will require a huge leg up. We know that food producers and food system labor is overrepresented in this group, so focusing on this issue will improve conditions for the people and communities that work hard to supply our food and also depend the most on the sustainable use of natural resources for their own livelihoods.

As we have already seen, the caloric conversion of plants to meat – especially that of mammals – requires huge inputs of land, water, and fertilizers. Not only that, but the result of this process pollutes waterways, kills oceanic life, and renders land unproductive. Who suffers the most? Yes, the poorest in society, who are often those living downstream from the rich. The world cannot hope to achieve Goal 10 without a concerted and coordinated effort to encourage a key shift in our dietary preferences.

GOAL 11: SUSTAINABLE CITIES AND COMMUNITIES

For better or for worse, more than half of the human population now lives in cities. By 2050, the number of urbanites will have risen to 6.5 billion people or two-thirds of all humanity. Extreme poverty is also often concentrated in these urban spaces. Making cities safe and sustainable will necessitate ensuring access to safe and affordable housing and upgrading all slum settlements. It will also entail investing more in public transportation, creating green public spaces, improving urban planning, and shifting our diets away from mammal meat.

Cities need to be prepared to support their growing, young populations and secure their needs, not the least of which is for healthy food. Cities don't produce much food; they rely on rural areas to sustain their people. City administrations thus could do a lot more to manage their food supply systems sustainably. Indeed, the Stockholm-based EAT Commission advised municipal administrations to consider establishing zoning regulations that promote urban agriculture and community gardens as a way to produce food locally and also work to support biodiversity and ecosystem services.

Further still, local farmers and producers in urban spaces could be better supported with incentives to produce healthy foods by facilitating market access, and shortening supply chains. The World Health Organization believes green spaces such as parks and sports fields, woods and natural meadows, wetlands or other ecosystems, represent a fundamental component of any urban ecosystem. Green urban areas facilitate physical activity and relaxation, and form a refuge from noise. Trees and plants produce oxygen, and help filter out harmful air pollution, including airborne particulate matter. Water spots, from lakes to rivers and fountains, help to moderate temperatures.

What has not been adequately explored in many cities, however, would be the extent to which some of these green spaces could be used to produce the food that people in the cities eat. There are, of course, unique challenges to urban agriculture – often a lack of space and securing enough sunlight amid the high-rise buildings of city environments. However, with innovative solutions, political will and community dedication, it is entirely possible to grow fruit and vegetables and even keep bees in urban areas. Indeed, in 2011 San Francisco passed city zoning legislation to allow agricultural activity in all areas of the city and redefined the parameters by which producers could sell their goods, making the local economy for small-scale producers boom. Many Bay area restaurants now sell meals made from food grown within San Francisco city limits.

GOAL 12: RESPONSIBLE CONSUMPTION AND PRODUCTION

According to leading demographers at the United Nations, more people globally are expected to join the middle class over the next two decades than ever before. This is good for individual prosperity, but it will increase demand for already constrained natural resources. If we don't act to change our consumption and production patterns, this shift will cause irreversible damage to our environment, thus erasing the gains and throwing millions of people back into poverty. We have a chance now to set things right. While we have never been so close to lifting everyone out of poverty, we have also never been so close to complete ecological collapse. The choice of direction is ours to make.

While we have never been so close to lifting everyone out of poverty, we have also never been so close to complete ecological collapse.

Achieving sustained and sustainable economic growth in line with the SDGs will necessitate the urgent reduction of our ecological footprint by changing the way we produce and consume both goods and resources. We also need to look at the incredible waste that is a by-product of the current scenario for these activities. One-third of all food produced is never even eaten by people – despite the fact that, as mentioned earlier, 815 million people are hungry, and every third person is malnourished. The impact of such loss and waste worldwide is tremendous. Food loss and waste is responsible annually for $940 billion in economic losses and nearly ten percent of greenhouse gas emissions.[124]

Goal 12 calls for halving a per capita global standard for food waste at the retail and consumer levels and reducing the food losses along the production and supply chains (including post-harvest losses) by 2030. It puts the onus on "every country, every major city, and every company involved in food supply chains" to set food loss and waste reduction targets that will ensure sufficient attention and a positive focus.

Agriculture is the biggest user of water worldwide, and irrigation now claims close to 70 percent of all freshwater for human use – indeed mostly to produce mammal meat for human consumption. Because there are so many ways to vary diets and food production, this is one area where it is possible to set clear and achievable targets for more responsible consumption and production and then drive related innovation and successful large-scale changes.

GOAL 13: CLIMATE ACTION

Today, 97 percent of climate scientists agree that human activity is creating a worldwide climate crisis . It is hard not to turn on the television, computer, or Smartphone these days without hearing discussion and statistics about climate change. Greenhouse gas emissions continue to rise to the extent that they are now more than 50 percent higher than their 1990 levels, when the mainstream media first noted the threat publicly.

Many scientists believe it is still possible, with political will and a wide new array of technological measures, to limit the increase in global mean temperature to two degrees Celsius above pre-industrial levels. Affordable, scalable solutions are becoming increasingly available to enable developing countries to leapfrog these issues and become cleaner, more resilient economies than those in the "industrialized" countries. The pace of change is also quickening, as more people are turning to renewable energy and other measures to reduce emissions and increase adaptation efforts.

Yet while climate change has entered the lexicon of daily conversation, not enough people are aware that their dietary choices are part of the problem. As mentioned earlier, producing just a kilogram of beef (two pounds) emits 26 kg (57 pounds) of carbon dioxide. The agriculture industry accounts for roughly 20 percent of the greenhouse gases that are responsible for climate change. Rethinking that burger thus can could make you a pioneer in problem-solving and a part of the solution.

GOAL 14: LIFE BELOW WATER

More than three billion people depend on marine and coastal biodiversity for their livelihood. According to the UNFAO, fish and fish products account for 17 percent of all animal protein consumed in the world, and 26 percent of that 17 percent is consumed in the poorest and least developed countries. The ocean also provides an important source of income for 60 million people who work in fisheries and aquaculture. However, nearly 90 percent of the world's marine fish stocks are now fully exploited, overexploited, and/or depleted.[125] One-third of the world's fish catch is also fed directly to livestock to be inefficiently converted into beef, thereby wasting significant amounts of this precious resource. If we are not careful, this overexploitation of "life below water" could push the regenerative capacity of the oceans past the point of no return.

90 percent of the world's marine fish stocks are now fully exploited, overexploited, and/or depleted.

As we have already seen, the oceans are suffering from aquatic and marine dead zones that are caused by an increase in nutrients in the water (particularly nitrogen and phosphorus). This process, known as eutrophication, sucks oxygen from the water and kills all life. In March of 2004, when the UN Environment Program published its first Global Environment Outlook Year Book (*GEO Year Book 2003*), it reported 146 dead zones in the world's oceans where marine life could not be supported due to depleted oxygen levels. Some of these zones were as small as a square kilometer (0.4 mi^2), but the largest, just off the coast of the United States, covered 70,000 square kilometers (27,000 mi^2). A 2008 study counted 405 dead zones worldwide.[126] In 2017, scientists recorded the largest dead zone off the Texas/Louisiana coast, where the rivers that flow down from the American heartland dump into the Gulf of Mexico.[127] Just upstream from these dead zones are the farms that feed our fixation for fast food Fatburgers

By reducing or avoiding mammal meat, you can directly contribute to the solutions necessary to restore the health of our oceans, restoring life below water to its balance within the ecosystem.

GOAL 15: LIFE ON LAND

We all know our lives depend on the health of the Earth for our sustenance and our livelihoods. At least 1.6 billion people depend on forests for their livelihood, and 75 percent of the world's poor are affected directly by land degradation. Plant life provides 80 percent of our human diet, and we rely on agriculture as an important economic resource. Forests account for 30 percent of the Earth's surface, and provide vital habitats for millions of species. Forests also are important sources of clean air and water, and are crucial for combating climate change through photosynthesis – the process through which plants convert CO2 from the air into biomass.

However, as highlighted above, today we are seeing unprecedented land degradation caused by human activity that includes extensive livestock production. The grazing of livestock takes up 26 percent of the Earth's terrestrial surface, and feed-crop production needs about a third of all arable land on Earth. Expansion of livestock grazing land is also a leading cause of deforestation, especially in Latin America. In the Amazon basin alone, roughly 70 percent of previously forested land is being used as pasture, while feed crops cover a large part of what remains.[128]

Goal 15 challenges the world to protect, restore and sustainably use terrestrial ecosystems, manage our forests, and halt and reverse land degradation and biodiversity loss. Eighty percent of endangered mammals are now threatened by habitat loss due to ever expanding agriculture that feed the few mammals that we eat. We simply must bring food production back within the limits of planetary boundaries. Reducing our consumption of mammal meat and eating a locally-based diet that is sustainably sourced, is -- in a nutshell-- the single biggest action we can take to protect life on land. And remember – life on land includes us.

GOAL 16: PEACE, JUSTICE, AND STRONG INSTITUTIONS

For humanity to move forward towards a place where it can realize its full potential, peaceful, just and inclusive societies are needed. It is well known that food insecurity and food price volatility can cause and/or exacerbate conflict, violence, and civil unrest. The war in Syria has been attributed, in part, to drought and rural distress following crop yield declines and food scarcities. The Arab Spring uprisings have been traced back to the 2011 price spikes for staples on international markets.

There is also a direct causal relationship between food on one side and peace, justice, and strong institutions on the other. A lack of any of the latter can be a root cause of dysfunctional food systems, hunger, and poor nutrition. Conversely, a problem with food systems, caused by failing crops drought or otherwise, can have a debilitating effect on the institutions that support peace and justice. Getting it ”right” with food can depend on this

16th Goal and strengthen opportunities to establish the conditions for societies to flourish. Rethinking that ribeye may seem trivial in this particular context, but believe me, it is not – for all the environmental reasons outlined in this *Manifesto*.

GOAL 17: PARTNERSHIPS FOR THE GOALS

In order to achieve these goals and ensure humankind charts a way forward that is sustainable and equitable, we must realize that we are all in this together. Personal choices have repercussions far beyond one's personal space, either hurting or healing the world at large. It's all about partnership and community— looking after one another. The partnerships to support the realization of these 17 Sustainable Development Goals must of course happen at the global level of nation states, but it is also required of subnational levels of government like states and provinces, and of cities, of communities, of clubs and associations, and of you, the reader of this *Manifesto*.

Personal choices have repercussions far beyond one's personal space, either hurting or healing the world.

Food offers an excellent arena for building connections, collaborations, and strong partnerships, precisely because almost everything can be traced to food, and almost all social and planetary issues that these SDGs want to tackle do have a food component. Food is an incredible umbrella that farmers, young people, scientists, business, inventors, politicians, indigenous peoples, workers, creatives, and pretty much everyone else can be brought together under to share solutions and craft a new vision for the future – a future of shared prosperity.

6: Conclusion: FINAL THOUGHTS AND SUGGESTIONS

AS ESTABLISHED IN THE INTRODUCTION, and what is already common knowledge to many, we human beings are omnivores. As omnivores, we have the ability to eat and survive on both plant and animal matter. In other words, we have a choice -- to kill or not to kill. And if we choose to kill, we can choose what, when and how to kill.

My hope with this *Manifesto* is to nudge a few people in the direction of choosing their food with more awareness – to question that part of our consumer culture that produces cellophane-wrapped sirloin steaks that are completely and consciously disconnected from the mammal that was. If veganism and vegetarianism are not for you, or if you are not yet ready to give up the occasional steak or hamburger, you can still be more conscious of the origin and impact of your dietary decisions, and as you realize the need to act, you can reduce the amount of mammal meat you consume until you are ready to give it up all together.

I have noticed in recent years that awareness of this issue is spreading. I am also encouraged that Americans have reduced their beef consumption by 19 percent from 2005 to 2014. Canadians are doing even better, now eating 25 percent less beef and pork since 1999.[129]

CONSIDER REDUCING YOUR MAMMAL MEALS

If you can't go "cold turkey" with the cow, perhaps try to reduce the amount of meat you do consume. In his 2009 book, *Food Rules: An Eater's Manual*, Michael Pollan says to consider changing the traditional portion sizes on your plate. Instead of an eight-ounce steak and a four-ounce portion of vegetables, serve four ounces of [grass fed] beef and eight ounces of veggies". In other words, treat meat as a flavoring or special-occasion food – not as the center attraction of every meal.

If you can't go "cold turkey" with the cow, perhaps try to reduce the amount of meat you do consume.

The January 2019 study undertaken by the EAT-Lancet Commission on Food, Planet and Health outlined the ideal healthy diet – one that is best both for the health of the individual and the planet. The authors of the study, who are international experts in health, nutrition, and sustainability, suggested that if you feel you cannot do without, then eat no more than a single hamburger each week. According to these researchers, a hamburger a week is about as much mammal meat as people should eat to improve their own health and the health of the planet.[130] Eating meat just once a week might be a challenge for some, but I believe it is a challenge worth taking on and overcoming.

A group of researchers who focus on heart disease pointed out in an article on the Harvard Medical School website that you don't need to become a vegetarian to protect your health, "but you can make modest improvements to your diet by gradually cutting back on the amount of meat you consume."[131] In a Harvard School of Public Health study cited in that article, substituting a healthier source of protein for just one daily serving of red meat lowered the participants' risk of dying prematurely. Switching to poultry, for example, dropped the risk by 14 percent, eating nuts lowered it by 19 percent, and substituting fish reduced it by seven percent. In other words, reducing or eliminating mammals from your diet can add healthier years to your life. Indeed, it's never too late to turn it around (but don't wait too long).

I am very encouraged by the number of friends who are starting to eat less mammal meat – often avoiding it altogether except for special occasions. Once every few months for the past ten years, a dozen of my closest friends get together for "Steak Night" in New York. They "meat" four times a year to have an evening of carnivorous chaos. The rest of the year they try to reduce their animal protein intake to those that are not mammals, like chicken, turkey or seafood.

VEGEBURGERS ARE THE NEW COOL

Veggie and vegan burgers are becoming increasingly popular across both the United States and Canada. Traditionally meat-centered restaurants like McDonald's, Red Robin, and The Cheesecake Factory now offer meatless sandwich options. In fact, in the 12 months prior to August 2018, plant-based meat sales skyrocketed worldwide by 24 percent, topping $670 million in sales in the United States alone, according to data from Nielson. [132] In the United Kingdom, similar trends are being seen. Within days of launching Vivera's vegan steak in May, the UK retail giant, Tesco, announced it had sold almost 40,000 units.[133]

Perhaps the New York Steak Night Club could eventually start meeting at the famous Manhattan restaurant called Momofuku Nishi, where renowned chef, David Chang, has a new item on the menu: The Impossible Burger. It looks like the real deal -- crispy brown on the outside, juicy pink on the inside – it sizzles on the grill and it smells like beef. It even bleeds.

No animals are killed to make the Impossible Burger. It is made entirely from vegetable products, including wheat, coconuts, and potatoes, but it also has a secret ingredient: heme, a molecule that makes meat taste delicious and which Impossible Foods, Inc., has recreated by fermenting yeast.

A veggie burger can actually be more filling and satisfying than beef burgers.

The global meat alternative market also is expanding-- a recent market forecast predicts it will reach $6.3 billion by 2023, from the current value of $4.63 billion.[134] The Los Angeles-based company, Beyond Meat, is taking advantage of this opportunity by taking on the $1.4 trillion meat industry. In 2018, it expanded to more than 35,000 restaurants, grocery stores, universities, hotels, and stadiums. It also built a 26,000-square-foot R&D lab dedicated to diversifying its product portfolio of non-meat food that looks and tastes like meat.

In March 2019, Beyond Meat announced Beyond Beef, a new product meant to taste, feel, and smell just like ground beef – but with 25 percent less saturated fat than beef. Their biophysicists studied the molecular structure of meat and identified the plants needed to replicate the texture and taste of beef. Beyond Beef is made from a blend of peas, mung bean, and rice proteins intended to rival the popular ingredient found in empanadas or tacos.[135] The taste is so close to the real deal that according to Beyond Meat Founder, Ethan Brown, 93 percent of people who now purchase the company's products in grocery stores are meat eaters.

However, Beyond Beef has not yet been able to compete with the beef industry on cost. Beyond Burger retails at $5.99 for two patties -- 71 percent more expensive than cow flesh when compared pound for pound. This issue may soon be resolved, as the company is looking into a variety of plant proteins like sunflower seed, mustard seed and lupin to help lower the price.[136] If and when they are able to sell their product at a lower price point than beef, but with better taste and healthier results, expect Beyond Beef's share of the burger market to expand rapidly. Smart money is probably moving in that direction already. Impossible Foods (and its Impossible Burger) easily raised US$182 million from investors that include Bill Gates and Google Ventures to make the widespread availability of their burger possible.[137]

BUT ARE THESE NEW FOODS SATISFYING?

Many meat-eaters would scoff at the idea of a veggie burger satisfying a man's hunger after a hard day's work. However, a recent blind study would prove them wrong – a veggie burger can actually be more filling and satisfying than beef burgers. A study published in the journal, *Nutrients,* in December 2018 reported that men – when unaware of which kind of burger they were eating, overwhelmingly reported feeling fuller after eating a vegan burger. The scientists studied 60 men — 20 healthy, 20 obese, and 20 with Type 2 diabetes. All had their physiological responses recorded after each meal.

Researchers were baffled by the results until they realized that vegan meals actually produce higher levels of beneficial gastrointestinal hormones that affect the regulation of glucose metabolism, energy homeostasis, satiety, and weight management.[138]

NEW BURGERS ON THE HORIZON

In late 2018, the Italian bioengineer, Giuseppe Scionti, from the Spanish start-up company, Novameat, unveiled the world's first 3D-printed meat-free steak made from vegetable proteins. This "fake steak" is made from vegan rice, peas, and seaweed and still provides the amino acids needed for a healthy diet. They are turned into a food paste that is 3D-printed to form a raw, steak-like substance that can be fried, grilled on the oven, or thrown on a barbeque. Scionti's method has allowed him to print a 100-gram piece of steak in around 30 minutes.

I personally do not feel the need to fool my taste buds into thinking I am eating a pig or a cow. I am fine with eating a steak that might taste a bit like vegetables – or a turkey burger that tastes like turkey. However, given the assumption that certain people will never give up their preference for eating mammals, all this new research is worthwhile in my opinion.

EXPANDING OUR HORIZONS

Our attitudes towards diet are a fundamental part of our "culture". These attitudes have been processed, legitimized and transmitted over generations, shaping our habits and our identities. At the same time, scientific inquiry and new information can greatly contribute to further introspection and allow us to question certain long-presumed assumptions that we may have held concerning diet. Achieving any transformation can be both complicated and complex and even emotional sometimes.

Crops that have been long forgotten over the last century are now being rediscovered.

By and large, our current diet is based on a very narrow range of food options. Of the 30,000 available and edible plant species on Earth, very few are actually consumed. Yet these neglected and underutilized crops can help increase diversification in our systems of food production. New or rediscovered crops could add fresh and exciting items to our menus that offer a better supply of nutrients, essential amino acids, fiber, and new proteins. These new products could have more positive impacts on the environment and our health and even our future as a species.

Fortunately, crops that have been long forgotten over the last century are now being rediscovered. Traditional crops are often very nutritious and offer us a more well-balanced diet. Quinoa, for example, is the only cereal that contains all the amino acids needed by humans. The bambara ground nut is another item that offers us a great source of protein, and millet is high in calcium and iron. The search must go on for other sources of sustenance. We should not limit our opportunities when it comes to food – expanding the boundaries of our imagination could yield new surprises that are much better than what we are hitherto used to. We just need to branch out and explore a little.

BUG BURGERS?

In April 2019, Abdulrahman Hassaballah, a PhD candidate at New York State University at Buffalo, won $3,000 in a sustainability competition by proposing the world shift to insect burgers from those made from beef. "For every 100 beef burgers we eat, that's around 750 pounds of greenhouse gas emissions," he said upon receiving his award, "For every 100 insect burgers, only 10 pounds of CO2 is produced."[139]

Hassaballah is not far off the mark, if only he can get western consumers to adopt what is commonly accepted in many parts of the world. Around two billion people regularly eat insects as part of their daily diet, and over 1,900 species are edible. The most commonly eaten bugs are beetles, caterpillars, bees, wasps and ants, and they are on the menu in many parts of Asia, Africa and Latin America.

In the United States, we have been eating bugs for decades without even knowing it. The Defect Levels Handbook published by the U.S. Food and Drug Administration outlines legal levels for insects in our food.[140] Take beer for example—the acceptable limit of insect infestation in hops is 2,500 aphids (small sap-sucking insects like the blackfly) per 10 grams. Canned fruit juices are allowed up to one maggot per 250 ml, curry powder is allowed up to 100 insect fragments (head, body, legs) per 25 grams and chopped dates are allowed up to 10 whole dead insects. The list goes on and on. Therefore, shifting our burgers to bugs should in theory not bother us too much because we have already been eating them for years.

In 2017, the Swiss supermarket chain Coop began selling burgers made with mealworms, rice and vegetables, flavored with oregano and chili. Essento, the Swiss start-up company that launched the same year, claims business is booming and that it is having a hard time keeping up with demand. Its three best-selling products are Essento Insect Burgers, Essento Insect Balls and Essento Grasshoppers skewers.[141] The Swiss, which are well known for the quality of their range-fed beef and their wild game meat now seem to be ahead of the curve in Europe when it comes to eating insects as a meat alternative.

THE SEARCH FOR ARTIFICIAL MEAT

In the 2009 animated film, *Cloudy With a Chance of Meatballs*, Flint Lockwood invents a machine that transforms water into food. This machine, which he calls the Flint Lockwood Diatonic Super Mutating Dynamic Food Replicator (FLDSMDFR), uses the principle of "hydrogenetic mutation" wherein water molecules are "bombarded with microwave radiation that mutates their genetic recipe into any food you want."

While obviously this is science fiction, something similar from actual science is now not too far off. Scientists across the globe are working on a synthetic solution to the demand for meat. Artificially -produced meat looks like meat, feels like meat, and is meat for all intents and purposes. However, this artificial or "cultured" meat is grown from stem cells in giant vats. The technology still has a long way to go before it will be served regularly at Black Angus Steakhouse, but it is moving in that direction, and its time will come.

The creation of this meat – known as "clean meat" in the industry – starts with a small amount of muscle tissue from an animal you wish to replicate. Technicians collect stem cells from this tissue, multiply them, and allow them to differentiate into primitive fibers that then bulk up to form muscle tissue. A spokesperson from Mosa Meat, one of the companies currently investing in "clean meat" research and development, says that one tissue sample from a cow can yield enough muscle tissue to make 80,000 quarter-pounders. Someday this technology will be able to serve up lab-grown beef, pork, poultry, and seafood at your family restaurant – producing it on demand.

In early 2019, researchers at the University of Bath in the United Kingdom succeeded in growing meat cells on blades of grass. Their process uses grass as structural support (known in the industry as "scaffolding") to allow stem cells from pigs to proliferate into edible meat. The team has successfully grown pig cells using its grass scaffolding, which opens the door to growing bacon in a lab-setting – and ultimately in a commercial setting. Whether or not such artificially-grown meat will be better for your health than the real meat, remains to be seen. But at least it will be better for the planet.

Much more research and development is needed of course until lab-grown clean meat will replace the need to raise billions of cows and pigs for slaughter. However, if the capital markets are any indication of these future developments, there is considerable promise on the horizon. In 2017 alone, just one of these companies in this space, Memphis Meats, took in $17 million from sources that included Bill Gates and the agricultural company, Cargill.[142]

THE NOT-SO-SECRET LIFE OF PETS

It is obviously not their fault, but our pets are also part of the problem – and conversely, they can be a big part of the solution. What we are giving them to eat is exacerbating an already dangerous situation, especially as pet ownership increases in many of the developing countries like China and India.

On the health front, there is no reason not to believe that all of the risks associated with humans eating other mammals is not also the same for our cats and dogs. Also, with respect to the environment, and as mentioned already, total meat consumption by domestic pets is responsible for the emission of 64 million tons of carbon dioxide a year – around the same climate impact as driving 13.6 million cars.[143]

If this *Manifesto* resonates with you, it is not too difficult to bring your pets along with you on your journey so they too can avoid eating their fellow mammals, reduce their carbon footprint and safeguard the environment for future generations of felines and canines. All you have to do is read the label of your pet food before you purchase it.

The website for the Association of Feed Control Officials (AAFCO) explains how to read the ingredients on a pet food container: [144]

> *"Meat is the clean flesh derived from slaughtered mammals and is limited to that part of the striate muscle which is skeletal or that part which is found in the tongue, in the diaphragm, in the heart or in the esophagus; with or without the accompanying and overlying fat and portions of the skin, sinew, nerve, and blood vessels which normally accompany the flesh. It shall be suitable for animal food. If it bears a name descriptive of its kind, it must correspond thereto."*

In other words, "meat" is primarily the muscle tissue of the animal, but it may also include the fat, gristle and other tissues that normally accompany the muscle, like what you see in a portion of raw meat sold for human consumption. In addition to using the term "meat," the pet food manufacturer may also identify the species from where the meat was derived, such as "beef" or "pork." However, to use the generic term "meat" on the label means the food can only come from cattle, pigs, sheep, or goats. If it comes from any other mammal, that species must be identified (for example, as with "buffalo" or "venison").

If the muscle is from a non-mammalian species, such as poultry or fish, it cannot be declared as "meat", and the appropriate identifying terms must be used. This description is what I would recommend you look for if you agree with the principles of this *Manifesto* and wish to include your pets in your efforts to be part of the solution.

SUPPORT SOUND GOVERNMENT POLICIES

Our individual consumer choices may not be enough to avert what the January 2019 Lancet report calls "catastrophic damage to the planet." Governments also will need to encourage healthy food choices and ensure access to nutritious food. Policies and government subsidies will need to be redirected away from harmful agricultural practices and toward ones that are healthier for our bodies and our environment and indeed our planet.

It is clear that the societal costs of mammal meat consumption are far greater than the price paid by the consumer. There is now increased discussion in many places around the need to formulate a governmental policy response that regulates red and processed meat consumption similar to the regulations for other carcinogens and foods with public health concerns. One approach is to regulate the industry or to outlaw certain foods – as New York City has tried to do by banning sugar-sweetened drinks in cups larger than 16 ounces (0.5 liters). A more market-based approach would involve taxing red and processed meat according to their health impacts. This latter approach looks at the cost of eating meat on the global economy and how much tax people should pay to offset the health and environmental consequences of their diets.

Looking into the most optimal taxation levels for red and processed meats in nearly 150 countries and regions, health experts at Oxford university concluded in its 2018 study that introducing a tax on meat would produce widespread health and environmental benefits. In high-income countries, the price for beef, lamb, and pork would need to be increased by more than 20 percent, while processed meats like sausages and hot dogs would need to more than double in price to cover their true cost to society. These researchers concluded that introducing such a health tax on these products would offset healthcare costs and likely prevent more than 220,000 deaths a year globally.[145]

Some argue that if the true cost of meat production were reflected in the price of the meat itself, then only elites would be able to eat meat. If so, so be it. There are many things that are so expensive that only the rich can afford them. Just because private jets can be afforded by the super wealthy doesn't mean we should argue that they should be subsidized for everyone else. Let the rich eat their expensive beef that reflects the products true cost, while the rest of us eat more healthy alternatives.

A TRANSFORMATIVE CHANGE IS NEEDED IN SOCIETY AND IN OUR ECONOMIES

Achieving the vision outlined in this book will obviously require a dramatic and transformative shift within both our society and the economy at large. In the United States, the mammal meat industry is responsible for 5.4 million jobs and $257 billion in wages. An estimated 527,019 people have jobs in production and packing, import operations, sales, packaging and the direct distribution of mammal meat products. One report claims the meat industry accounts for $1.02 trillion in total economic output or in other words, 5.6 percent of gross domestic product (GDP) – in the United States alone. [146]

Livestock also plays a crucial economic role for an estimated 60 percent of rural households in developing countries – including small-holder farmers, agro-pastoralists and pastoralists. It contributes to the livelihoods of about 1.7 billion poor people. According to the UNFAO, at least 70 percent of those employed in the sector are women.[147] Livestock, including dairy and other animal products, creates cash and in-kind incomes and enables savings for future needs. As a result, it should be clearly recognized here that this sector – while causing harm to our health, society and the environment-- also plays a major role in reducing poverty.

Achieving the vision outlined in this book will obviously require a dramatic and transformative shift within both our society and the economy at large.

If everyone were to stop eating mammal meat immediately after reading this book, it would probably push a lot of people into poverty. However, as with all disruptive technologies, shifts in the market require economic adaptation. The suppliers and supply-chain management infrastructure will resist like they always do – but eventually they will have to adapt, people will need to be retrained, and new jobs will need to be created. What is needed is a more sustainable alternative which can offer new technologies and thus new jobs, and an accompanying shift away from relying on the exploitation of our fellow mammals to fuel the economy.

Looking ahead, I do believe the path to inclusive prosperity will include a dramatic reduction in the production and consumption of our fellow mammals. Overcoming the complex challenges that the world is now confronting will require a political willingness to embrace the principles of sustainability and transformative action to tackle the root causes of poverty and hunger successfully. What we choose to grow, kill and eat will play a big role in this effort.

HOLD THE CHEESE, PLEASE – IF IT HAS RENNET

One of the most difficult processes for me in the transition away from mammal meat was discovering from a vegetarian roommate that most cheese contains animal rennet, a traditional cheese-making ingredient that is extracted from the stomach lining of slaughtered cows, goats, sheep, and even pigs. Rennet allows these young animals to digest the mother's milk. This enzyme curdles the milk, separating curds from whey, thus completing the first step in the process of making cheese.

I was living in Switzerland when I learned this, and Europe is famous for its many delicious varieties of cheese. I was shocked to learn that most European cheeses are made with animal rennet because their age-old recipes have always used it – and are actually required to by law. For example, according to European Union law, Parmesan must contain animal rennet in order to be called Parmigiano-Reggiano.[148]

The path to inclusive prosperity will include a dramatic reduction in the production and consumption of our fellow mammals.

Rennet alternatives are necessary for the purposes and principles outlined in this book -- or even for those who simply don't like the idea of baby animal stomachs in their cheese. Commercial cheese makers do not have to reveal whether the source of their rennet enzyme is artificially synthesized, is real rennet gleaned from a stomach, or is a non-rennet alternative. Thus, when in doubt, look for cheeses marked "vegetarian" or "kosher". You can also ask a knowledgeable cheese monger for the names of cheeses made with veg-friendly alternatives, such as vegetable rennet. These typically come from thistle plants, while microbial rennet is derived from fungus, yeast, or mold. If you live in the United States, Trader Joe's and Whole Foods both label the sources of the rennet used in their generic-brand cheeses, many of which are indeed vegetarian.

SOME CAUSAL CONSIDERATIONS

While preparing for a ten-day Vipassana meditation retreat at Wat Suan Mokkh in Chaiya in Southern Thailand, I listened to an English-speaking monk lecture a group of us on the laws of causation. The monk explained that causation (or cause and effect), is ultimately the essence of karma. It is what connects one process (the cause) with another process or state (the effect), wherein the first is partly responsible for the second, and the second is partly dependent on the first. "The thing to remember," he said, "is good deeds bring good results; bad deeds bring bad results; your own deeds bring your own results."

Later that same day, the monk elaborated on the teachings of the Buddha about the karma of killing. "If you kill one sentient being," he explained, "then for 500 lifetimes you will be killed by others. This is the result of the negative karma of killing."

I raised my hand to ask a question, knowing that the following day when the retreat began, we would be forbidden from saying any word to anyone for ten days. So I asked:

> *"If the killing of any sentient being brings uncomfortable consequences to the killer for 500 lifetimes, then by the laws of causation, should one also look at the chain of events leading up to that killing to understand ultimate karmic responsibility. If the killing is the end result of a line of linked causal actions that can be determined, and if the primary linked action does not take place, then the killing would have been avoided. Therefore, it makes sense to me that the first action would be the one carrying the most karmic debt, and that individual who killed would thus be the one being killed by others 500 times."*
>
> *"In other words," I continued, "when I purchase a steak at the market, am I in fact ultimately responsible for its death – even if I am completely unaware of the cow?"*
>
> *"Yes, you are," replied the monk. "You are even more guilty than the slaughterhouse executioner. For if there were no market demand for beef, there would be no need for killing of the cow."*

I have often observed that the food the 300,000 monks in Thailand receive every day in their begging bowls consists of rice, vegetables and meat. As the people of Thailand revere and respect their monks, it stands to reason they would seek to model their own lives accordingly. When they see the holy monks receiving and eating meat, then in their view eating meat must be permitted, probably even encouraged.

As I set upon my ten days of mindful introspection and contemplation, I pondered the possibility that if the monks chose to not accept meat, and encouraged people to give them vegetables instead, then the number of animals being killed in that country would be dramatically reduced. Thus, with regards to the killing of those sentient beings that end up in their morning begging bowls, the karmic debt belongs to the monks, who will then have to suffer being killed themselves 500 times by others. Perhaps, I thought, those monks will return 500 times as cows and pigs, only to end up in the begging bowls and eaten by other monks in an endless cycle that could only end by the monks becoming vegetarians. An interesting thought to ponder.

AN AFTERWORD TO CONSIDER

If you are a committed carnivore and have made it to the end of this book, I would like to thank you first of all for having an open mind. If the arguments in this book make sense to you, please pass them on – and define your position as someone who does not eat mammals. When you say you don't eat red meat, you may find many people who will simply shrug and say "okay." But saying you "don't eat mammals" defines a position that usually inspires inquiry. When asked, do speak of the insights I shared with you in this book, as well as your own thoughts on these issues. Better yet, send them a copy of the *Manifesto* so they can read it for themselves.

If this book resonates, or even if it doesn't, I would love to hear from you with suggestions for the next edition Please join me at www.adamrogers.online, on LinkedIn, or on Twitter at @adamrogers2030.

ABOUT THE AUTHOR

ADAM ROGERS HAS BEEN EXPLORING the world and writing about life for much of the last 40 years. His explorations have taken him to more than 130 countries. He first began to question the links between diet, well-being, and the environment at the age of 15, after reading *Diet for a Small Planet*, the 1971 bestselling book written by Frances Moore Lappé. His first career effort was as a short order cook and then a sous chef at Charlies, the only French restaurant in Whitehorse, Yukon Territory, Canada, where he grew up. Despite being able to grill the best Chateaubriand in town, Adam became a vegetarian – which he later adapted to become a "no-mammalarian," which he outlines in this book.

Adam first started traveling in his late teens, paying particular attention to what people ate in different parts of the world. At the age of 18, he set off with a backpack and the intention to keep traveling east until he ended up back in the West. This peripatetic exploration of the planet took him to more than 50 countries on a budget under $100 a month. That experience formed the basis for his first book, *The Intrepid Traveler: The ultimate guide to responsible, ecological, and personal-growth travel and tourism*. He then went on to explore another 80 countries in different capacities as a writer, photographer, and as a development professional.

As the editor-in-chief of the Los Angeles-based *Earth News*, Adam covered the landmark 1992 Earth Summit in Rio de Janeiro and most of the subsequent summits during the 1990s in Barbados (Small Island Developing States), Cairo (Population), Istanbul (cities), Copenhagen (social development), and Beijing (empowerment of women). His book, *The Earth Summit, a Planetary Reckoning* documented the 1992 conference through the perspectives of youth, business, government, and civil society. Based on that experience, Adam wrote a book for the UN Environment Program, localizing the framework of Agenda 21 (*Taking Action: An Environmental Guide for You and Your Community*). Not long afterwards, Adam started working for the United Nations in various positions that lasted 22 years. Throughout this vital journey of discovery he has incessantly questioned what diet may have to do with the state of the environment and the prevalence of poverty in many parts of the world.

Adam received his BA in International Affairs from Northern Arizona University, his MA in Communication and Technology from the University of Alberta in Canada, and his MBA from the TRIUM program at New York University, the London School of Economics, and the HEC Paris School of Management. He is very active on social media, and can be found on Twitter, LinkedIn, Instagram and Facebook.

REFERENCES CITED

1 Fishbase, see https://www.fishbase.de/

2 Barrowclough GF, Cracraft J, Klicka J, Zink RM (2016, November 23) How Many Kinds of Birds Are There and Why Does It Matter? *PLoS ONE* 11(11): e0166307. Retrieved from https://doi.org/10.1371/journal.pone.0166307

3 Milman, O. (2018, 21 December) Why eating less meat is the best thing you can do for the planet in 2019. *The Guardian*. Retrieved from https://www.theguardian.com/environment/2018/dec/21/lifestyle-change-eat-less-meat-climate-change?CMP=fb_gu

4 Barton, RA; Harvey, PH (2000, June 29). "Mosaic evolution of brain structure in mammals". Nature. Retrieved from https://www.nature.com/articles/35016580

5 Aboitiz, F; Morales, D; Montiel, J (2003, October 26). "The evolutionary origin of the mammalian isocortex: Towards an integrated developmental and functional approach". Behavioral and Brain Sciences. Retrieved from https://www.ncbi.nlm.nih.gov/pubmed/15179935

6 Barton, RA; Harvey, PH (2000, June 29). "Mosaic evolution of brain structure in mammals". Nature. Retrieved from https://www.nature.com/articles/35016580

7 Freund, D. (2011, August 25). How animal welfare leads to better meat: A lesson from Spain. *The Atlantic*. Retrieved from https://www.theatlantic.com/health/archive/2011/08/how-animal-welfare-leads-to-better-meat-a-lesson-from-spain/244127/

8 Johnson, T. (2015, August 5). We used to count Black Americans as 3/5 of a person. For reparations, give them 5/3 of a vote. *The Washington Post*. Retrieved on (add dates for these sources for retrieval dates) from https://www.washingtonpost.com/posteverything/wp/2015/08/21/we-used-to-count-black-americans-as-35-of-a-person-instead-of-reparations-give-them-53-of-a-vote/?utm_term=.2ea4f793c1e6

9 Zolfagharifard, E. (2015, June 12). Move over Lassie: IQ tests reveal pigs can outsmart dogs and chimpanzees. *The Daily Mail*. Retrieved from http://www.dailymail.co.uk/sciencetech/article-3122303/Move-Lassie-IQ-tests-reveal-pigs-outsmart-dogs-chimpanzees.html

10 McElroy, D. (2002, January 6) Korean outrage as West tries to use World Cup to ban dog eating. *The Telegraph*. Retrieved from https://www.telegraph.co.uk/news/worldnews/europe/france/1380569/Korean-outrage-as-West-tries-to-use-World-Cup-to-ban-dog-eating.html

11 Bonhommeau, S. Dubroca, L. Le Pape, O. Barde, J. Kaplan, DM. Chassot, E. and Nieblas, AE. (2013, December 17) Eating up the world's food web and the human trophic level. *Proceedings of the Natural Academy of Sciences*. Retrieved from http://www.pnas.org/content/110/51/20617

12 Allen, N. (2009, October 19). McFarthest point in the US from a McDonald's. *The Telegraph*. Retrieved from https://www.telegraph.co.uk/news/newstopics/howaboutthat/6380193/McFarthest-point-in-the-US-from-a-McDonalds.html

13 Duprey, R. (2016, October 11). 15 Fascinating Things You Probably Didn't Know About McDonald's Corp. *The Motley Fool*. Retrieved from https://www.fool.com/investing/2016/10/11/15-fascinating-things-you-probably-didnt-know-abou.aspx

14 USDA (2018, January) Retrieved from http://usda.mannlib.cornell.edu/usda/current/LiveSlauSu/LiveSlauSu-04-18-2018.pdf

15 *Food Outlook: Biannual Report on Global Food Markets*, p. 7. FAO, October 2015. Retrieved from http://www.globalagriculture.org/fileadmin/files/weltagrarbericht/GlobalAgriculture/02Hunger/Oktober2015.pdf.

16 Okin GS (2017) Environmental impacts of food consumption by dogs and cats. *PLoS ONE 12(8): e0181301*. Retrieved 2019, January 13 from https://doi.org/10.1371/journal.pone.0181301

17 Willett, W., et al. (2019, January 16). Food in the Anthropocene: the EAT–Lancet Commission on healthy diets from sustainable food systems. *The Lancet*. Retrieved from https://eatforum.org/initiatives/the-eat-lancet-commission/eat-lancet-commission-summary-report/

18 Bird, J.K., Murphy R.A., Ciappio E.D., McBurney M.I. (2017, June 24) Risk of deficiency in multiple concurrent micronutrients in children and adults in the United States. *Nutrients*. Retrieved from https://www.ncbi.nlm.nih.gov/pubmed/28672791

19 Center for Disease Control and Prevention. (2016). Obesity and Overweight. Retrieved from https://www.cdc.gov/nchs/fastats/obesity-overweight.htm

20 Malamut, M. (2013, December 19). Eating Burgers From Restaurants Increases Obesity Risk, Shocking Study Says. *Boston Magazine*. Retrieved from https://www.bostonmagazine.com/health/2013/12/19/eating-burgers-restaurants-associated-higher-obesity-risk/

21 Swan SH., Liu F., Overstreet JW., Brazil C., Skakkebaek NE. (2007, June 22). Semen quality of fertile US males in relation to their mothers' beef consumption during pregnancy. Department of Obstetrics and

Gynecology, School of Medicine and Dentistry, University of Rochester. Retrieved from https://www.ncbi.nlm.nih.gov/pubmed/17392290

22 Lee, E. (2011, August 29). The Truth About Red Meat. *WebMD.* Retrieved from https://www.webmd.com/food-recipes/features/the-truth-about-red-meat#1

23 Harvard Medical School. (2012, January). What's the beef with meat? *Harvard Health Publishing.* Retrieved from https://www.health.harvard.edu/healthy-eating/whats-the-beef-with-meat

24 Skerrett, P. J. (2012, March 13). Study urges moderation in red meat intake. *Harvard Health Publishing.* Retrieved from https://www.health.harvard.edu/blog/study-urges-moderation-in-red-meat-intake-201203134490

25 Kolata, G. (2013, April 7). Culprit in Heart Disease Goes Beyond Meat's Fat. The New York Times. Retrieved from https://www.nytimes.com/2013/04/08/health/study-points-to-new-culprit-in-heart-disease.html?pagewanted=all

26 National Institutes of Health. (2019, January 18). Eating red meat daily triples heart disease-related chemical. Retrieved from https://www.nih.gov/news-events/nih-research-matters/eating-red-meat-daily-triples-heart-disease-related-chemical

27 Newman, T. (2018, June 15). Red meat allergy may increase heart disease risk. *Medical News Today.* Retrieved from https://www.medicalnewstoday.com/articles/322147.php

28 Bernstein, A., Pan, A, Rexrode, K.M., Stampfer, M., Hu, F.B., Mozaffarian, D. & Willett, W.C. (2011, December 29). Dietary Protein Sources and the Risk of Stroke in Men and Women. *Stroke.* Retrieved from https://www.ncbi.nlm.nih.gov/pmc/articles/PMC3288224/

29 Kimmell, J. (2018, August 2). The 5 behaviors that explain your risk of cancer. *Advisory Board.* Retrieved from https://www.advisory.com/daily-briefing/2018/08/02/cancer-5

30 Simon, S. (2015, October 26). World Health Organization Says Processed Meat Causes Cancer. Cancer.org / American Cancer Society. Retrieved from https://www.cancer.org/latest-news/world-health-organization-says-processed-meat-causes-cancer.html

31 American Institute for Cancer Research. (2009, July 22). AICR Statement: Hot Dogs and Cancer Risk. Retrieved from http://www.aicr.org/press/press-releases/aicr-statement-hot-dogs-and-cancer-risk.html

32 Micha, R., Peñalvo, J. Cudhea, F. (2017, March 7). Association Between Dietary Factors and Mortality From Heart Disease, Stroke, and Type 2 Diabetes in the United States. JAMA. Retrieved from https://jamanetwork.com/journals/jama/article-abstract/2608221

33 Virtanen, H., Koskinen, T., Voutilainen, S., Mursu, J., Tuomainen, T., Kokko, P., & Virtanen, J. (2017). Intake of different dietary proteins and risk of type 2 diabetes in men: The Kuopio Ischaemic Heart Disease Risk Factor Study. British Journal of Nutrition. Retrieved from https://www.cambridge.org/core/journals/british-journal-of-nutrition/article/intake-of-different-dietary-proteins-and-risk-of-type-

2-diabetes-in-men-the-kuopio-ischaemic-heart-disease-risk-factor-study/1929E2CC34B0A504B3A045FAA0569CA3

34 Shaw, J. (2012, January). A Diabetes Link to Meat. *Harvard Magazine*. Retrieved from https://www.harvardmagazine.com/2012/01/a-diabetes-link-to-meat

35 TriStar Medical Group, Total Women's Care. Vitamin B12. Retrieved from https://totalwomenscaretn.com/hl/?/22440/Vitamin-B12

36 Consumer Reports. (2012, November 9). Press Release: The Overuse of Antibiotics in Food Animals Threatens Public Health. Retrieved from https://advocacy.consumerreports.org/press_release/the-overuse-of-antibiotics-in-food-animals-threatens-public-health-2/

37 American Farm Bureau Federation. (2012, June 12) Letter to Congressperson Louise Slaughter. Retrieved from https://www.ahi.org/wp-content/uploads/2012/06/Congresswoman-Slaughter-science-June-12-final.pdf

38 Hoelzer, K. (2017, November 10). Answers to Common Questions About the Use of Antibiotics in Animal Agriculture. Pew Trusts. Retrieved from https://www.pewtrusts.org/en/research-and-analysis/articles/2017/11/answers-to-common-questions-about-the-use-of-antibiotics-in-animal-agriculture

39 United States Government, House of Representatives, Committee on Energy and Commerce. (2010, July 14). *Hearing: Antiobiotic Resistance and the Use of Antibiotics in Animal Agriculture, Subcommittee on Health, Energy and Commerce Committee, U.S. House of Representatives.* Retrieved from https://www.govinfo.gov/content/pkg/CHRG-111hhrg77921/pdf/CHRG-111hhrg77921.pdf

40 World Health Organization (2017, November). Antimicrobial resistance in the food chain. Retrieved from https://www.who.int/foodsafety/areas_work/antimicrobial-resistance/amrfoodchain/en/

41 Meat and greens: How bad for the planet is eating meat? (2013, December 31). *The Economist*. Retrieved from https://www.economist.com/feast-and-famine/2013/12/31/meat-and-greens.

42 Casalone, C; Hope, J (2018). Atypical and classic bovine spongiform encephalopathy. *Handbook of Clinical Neurology*. 153. pp. 121–134. ISBN 9780444639455.

43 North American Meat Institute (NAMI). (2017). The United States Meat Industry at a Glance. Retrieved from https://www.meatinstitute.org/index.php?ht=d/sp/i/47465/pid/47465

44 Fortune, A. (2017, September 11). US Meat Industry Debunks Cancer Link. Global Meat News. Retrieved from https://www.globalmeatnews.com/Article/2017/09/12/US-meat-industry-debunks-cancer-link

45 North American Meat Institute (NAMI). (2017). The United States Meat Industry at a Glance. Retrieved from https://www.meatinstitute.org/index.php?ht=d/sp/i/47465/pid/47465

46 Shike, J. (2019,January 16). Pork Industry Calls EAT-Lancet Report Radical, Irresponsible. *Farm Journal*.

Retrieved from https://www.agweb.com/article/pork-industry-calls-eat-lancet-report-radical-irresponsible/

47 Brower, M., and Warren L. (March 30, 1999). The consumer's guide to effective environmental choices: Practical advice. *The Union of Concerned Scientists*. Three Rivers Press; 1st edition.

48 Springmann, M., Clark, M., Mason-D'Croz, D., Wiebe, K., Bodirsky, B. L., Lassaletta, L.,... (2018, October 25). Options for keeping the food system within environmental limits. *Nature*. Retrieved from https://www.nature.com/articles/s41586-018-0594-0

49 Meat and greens: How bad for the planet is eating meat? (2013, December 31). *The Economist*. Retrieved from https://www.economist.com/feast-and-famine/2013/12/31/meat-and-greens.

50 Rossi, M. (2018, July 31). The Chinese Are Eating More Meat Than Ever Before and the Planet Can't Keep Up. *Mother Jones*. Retrieved from https://www.motherjones.com/environment/2018/07/the-chinese-are-eating-more-meat-than-ever-before-and-the-planet-cant-keep-up/

51 United Nations Food and Agriculture Organization. (2009, September 23). 2050: A third more mouths to feed. Retrieved from http://www.fao.org/news/story/en/item/35571/icode/

52 Lindsey, R. (2018, August 1). Climate Change: Atmospheric Carbon Dioxide. NOAA. Retrieved from https://www.climate.gov/news-features/understanding-climate/climate-change-atmospheric-carbon-dioxide

53 IPCC 2007, Summary for Policymakers, in Climate Change 2007: Impacts, Adaptation and Vulnerability. Contribution of Working Group II to the Fourth Assessment Report of the Intergovernmental Panel on Climate Change, Cambridge University Press, Cambridge, UK, p. 17.0*****

54 Milman, O. (2018, 21 December) Why eating less meat is the best thing you can do for the planet in 2019. *The Guardian*. Retrieved from https://www.theguardian.com/environment/2018/dec/21/lifestyle-change-eat-less-meat-climate-change?CMP=fb_gu

55 Heller, M.C. and G.A. Keoleian. 2014. Greenhouse gas emission estimates of U.S. dietary choices and food loss. *Journal of Industrial Ecology: Supporting Information*. Retrieved from https://doi.org/10.1111/jiec.12174

56 Fenelli, D. (2007, July 18). Meat is murder on the environment. New Scientist. Retrieved from https://www.newscientist.com/article/mg19526134-500-meat-is-murder-on-the-environment/

57 Harvey, F. (2016, March 21). Eat less meat to avoid dangerous global warming, scientists say. *The Guardian*. Retrieved from https://www.theguardian.com/environment/2016/mar/21/eat-less-meat-vegetarianism-dangerous-global-warming

58 Gerber, P.J., Steinfeld, H., Henderson, B., Mottet, A., Opio, C., Dijkman, J.... (2013). Tackling climate change through livestock – A global assessment of emissions and mitigation opportunities. *Food and Agriculture Organization of the United Nations (FAO)*, Rome.

59 Ibid.

60 Strom, S. (2017, March 21). Americans Ate 19 percent Less Beef From '05 to '14, Report Says. *The New York Times*. Retrieved from https://www.nytimes.com/2017/03/21/dining/beef-consumption-emissions.html

61 Kunzig, R. (2014, January). Carnivore's dilemma. *National Geographic Society*, 2014. Retrieved from http://www.nationalgeographic.com/foodfeatures/meat/.

62 Ibid.

63 Union of Concerned Scientists. (201). What's Driving Deforestation? Retrieved from https://www.ucsusa.org/global-warming/stop-deforestation/whats-driving-deforestation

64 Haan, C. & Blackburn H. (1997). Livestock-Environment Interactions: Issues and Options. United Nations Food and Agriculture Organization. Retrieved from http://www.fao.org/ag/againfo/resources/documents/Lxehtml/

65 Williams, A. (2019, January 19). Brazil closes 2018 with largest-ever beef volume exports. Global Meat News Retrieved from https://www.globalmeatnews.com/Article/2019/01/22/Largest-ever-beef-exports-by-volume-for-Brazil

66 Measuring the Daily Destruction of the World's Rainforests *Scientific American*, 2009. Retrieved from https://www.scientificamerican.com/article/earth-talks-daily-destruction/

67 Alexandratos N and Bruinsma J. (2012). *UN Food and Agriculture Organization*. World Agriculture Towards 2030/2050: The 2012 Revision No. 12-03. Retrieved from http://www.fao.org/fileadmin/templates/esa/Global_persepctives/world_ag_2030_50_2012_rev.pdf

68 Foley J A et al (2011, October 12). Solutions for a cultivated planet. *Nature*. Retrieved from https://www.nature.com/articles/nature10452.

69 Thompson, P. & Kituyi, M. (2018, July 13). 90 percent of fish stocks are used up – fisheries subsidies must stop. *World Economic Forum*. Retrieved from https://www.weforum.org/agenda/2018/07/fish-stocks-are-used-up-fisheries-subsidies-must-stop/

70 Alder, J., B. Campbell, V. Karpouzi, K. Kaschner and D. Pauly (2008, August). Forage fish: from ecosystems to markets. *Annual Review of Environment and Resources, 33, 153-166*. Retrieved from https://www.annualreviews.org/doi/10.1146/annurev.environ.33.020807.143204

71 Sustainable Fisheries (2019). How many Fisheries are Overfished? Retrieved from https://sustainablefisheries-uw.org/fact-check/how-many-fisheries-are-overfished/

72 Natural Resources Defense Council (2017, March), Issue paper: Less beef, Less carbon: Americans shrink their diet-related carbon footprint by 10 percent between 2005 and 2014. Retrieved from https://www.nrdc.org/sites/default/files/less-beef-less-carbon-ip.pdf

73 Strom, S. (2017, March 21). Americans Ate 19 percent Less Beef From '05 to '14, Report Says. *The New York Times*. Retrieved from https://www.nytimes.com/2017/03/21/dining/beef-consumption-emissions.html

74 Mintel. (2018, February 15). Press Release Taste is the Top Reason US Consumers Eat Plant-Based Proteins. Retrieved from https://www.mintel.com/press-centre/food-and-drink/taste-is-the-top-reason-us-consumers-eat-plant-based-proteins

75 Coffman, V. Milburn, D. (2018, September 12). Two-thirds of U.S. consumers say they are eating less meat. HUB/Johns Hopkins University. Retrieved from https://hub.jhu.edu/2018/09/12/consumers-cut-back-on-meat-consumption/

76 Davis, L.C. (2016, March 28). *The economic case for world-wide vegetarianism. The Atlantic.* Retrieved from https://www.theatlantic.com/business/archive/2016/03/the-economic-case-for-worldwide-vegetarianism/475524/

77 Ibid.

78 Hallock, B. (2014, January 27). To make a burger, first you need 660 gallons of water. The Los Angeles Times. Retrieved from https://www.latimes.com/food/dailydish/la-dd-gallons-of-water-to-make-a-burger-20140124-story.html

79 Park, A. (2013, December 19). 5 Surprising Things We Feed Cows. Mother Jones. Retrieved from https://www.motherjones.com/food/2013/12/cow-feed-chicken-poop-candy-sawdust/

80 Bittman, M. (2014, July 15). The True Cost of a Burger. *The New York Times*. Retrieved from https://www.nytimes.com/2014/07/16/opinion/the-true-cost-of-a-burger.html

81 Cawley J, Meyerhoefer C. (2012, January 31).The medical care costs of obesity: an instrumental variables approach. J Health Econ. 31:219-30. Retrieved from https://www.ncbi.nlm.nih.gov/pubmed/22094013

82 Springmanna, M, Godfraya, C, Raynera, M & Scarborough, P; Analysis and valuation of the health and climate change cobenefits of dietary change. *Proceedings of the National Academy of Sciences of the USA (PNAS)*, 2015, January 22. Retrieved 16 January 2019 from https://www.pnas.org/content/pnas/early/2016/03/16/1523119113.full.pdf

83 Schechinger, A. W. (2018, October 2). America's Nitrate Habit Is Costly and Dangerous. Retrieved from https://www.ewg.org/research/nitratecost/

84 City of Des Moines Water Works (2015, December 10). Nitrate Removal Facility Fact Sheet. Retrieved from http://www.dmww.com/upl/documents/water-quality/lab-reports/fact-sheets/nitrate-removal-facility.pdf

85 May, J. (2017, July 11) City of Hiawatha Issues High Nitrate Warning. *Hiawatha World*, Retrieved from www.hiawathaworldonline.com/news/city-of-hiawatha-issues-high-nitrate-warning/article_71208a3a-b98e-54a4-bc20-1c4b9bdd7a10.html

86 Vivian B. Jensen et al., (2012, July). Technical Report 6: Drinking Water Treatment for Nitrate. Center for Watershed Sciences, University of California, Davis. Retrieved from groundwaternitrate.ucdavis.edu/files/139107.pdf

87 Institute of Mechanical Engineers. (2013, November 2). Global food: Waste not, Want not. Retrieved from https://www.imeche.org/policy-and-press/reports/detail/global-food-waste-not-want-not

88 UN Food and Agriculture Organization. Key facts on food loss and waste you should know! Retrieved from http://www.fao.org/save-food/resources/keyfindings/en/

89 Lovgren, S. (2005, August 31). Chimps, Humans 96 percent the Same, Gene Study Finds. *National Geographic*. Retrieved from https://news.nationalgeographic.com/news/2005/08/chimps-humans-96- percent-the-same-gene-study-finds/

90 Rosemary, J. (2013, November 30). Pigs and Chimpanzees Living in Sin? *Psychology Today*. Retrieved from https://www.psychologytoday.com/us/blog/what-makes-us-human/201311/pigs-and-chimpanzees-living-in-sin

91 *Pigs: The Missing Evolutionary Human Link*. (2016, December 13). Retrieved 20 November 2018 from https://www.globalfounders.london/blog/pigs-the-missing-evolutionary-human-link

92 Cahill, M. (2016, January 26). Humans Are Genetically Related To Some Pretty Bizarre Things...So Weird. ViralNova. Retrieved from http://www.viralnova.com/genetic-relatives/

93 Steenhuysen, J. (2009, April 23). Cattle genes may give clues about human health. *Reuters*. Retrieved from https://www.reuters.com/article/us-cattle-genes/cattle-genes-may-give-clues-about-human-health-idUSTRE53M5XT20090423?irpc=914

94 The Editors of Encyclopaedia Britannica. Maori – Maori Culture in The 21st Century. *Encyclopædia Britannica*.

95 Eskimo Kisses, Arm Hair, Moon Flags & Spike Lee vs. Stan Lee vs. Bruce Lee. *Esquire Magazine*. 2007-05-09.

96 Coren, Stanley (2004, January 10). Why Dogs Touch Noses: Communication and More. *Psychology Today*. Retrieved 8 January 2019 from https://www.psychologytoday.com/us/blog/canine-corner/201001/why-dogs-touch-noses-communication-and-more

97 Zolfagharifard, E. (2015, June 12). Move over Lassie: IQ tests reveal pigs can outsmart dogs and chimpanzees. *The Daily Mail*. Retrieved from http://www.dailymail.co.uk/sciencetech/article-3122303/Move-Lassie-IQ-tests-reveal-pigs-outsmart-dogs-chimpanzees.html

98 Fuoco, M. A. (2002, April 9). LuLu the heroic pig now known worldwide. *Pittsburg Post-Gazette*. Retrieved from http://old.post-gazette.com/neigh_west/20020409lulu0409p1.asp

99 Price, M. (2019, January 25). 3-year-old boy lost in NC woods tells his family he 'hung out with a bear for 2 days'. *The Charlotte Observer*. Retrieved from https://www.charlotteobserver.com/news/local/article225090590.html

100 Kandlbinder, S. G. (2009) Saved By Dolphins: The Todd Endris Story. AuthorHouse.

101 Adams, W.L. (2010, November 12). Dick Van Dyke: Porpoises Saved My Life. *Time*. Retrieved from http://newsfeed.time.com/2010/11/12/dick-van-dyke-porpoises-saved-my-life/

102 Elliott, K. (1996, August 4). True tale or great white lie? *The Independent*. Retrieved from https://www.independent.co.uk/sport/true-tale-or-great-white-lie-1308156.html

103 Thomson, A. (2004, November 25). Dolphins saved us from shark, life-guards say. *New Zealand Herald*. Retrieved from https://www.nzherald.co.nz/nz/news/article.cfm?c_id=1&objectid=3613343

104 The Telegraph (2004, November 24). Dolphins 'save' swimmers from shark attack. Retrieved from https://www.telegraph.co.uk/news/1477378/Dolphins-save-swimmers-from-shark-attack.html

105 Chicago Tribune (2016, August 16).20 years ago today: Brookfield Zoo gorilla helps boy who fell into habitat. Retrieved from http://www.chicagotribune.com/news/ct-gorilla-saves-boy-brookfield-zoo-anniversary-20160815-story.html

106 Burrows, T. (2016, January 20). EXCLUSIVE - My guardian gorilla angel: 30 years on from the amazing moment ape saved tiny boy who fell into his enclosure, the man protected by gentle giant reveals how the incident has shaped his entire life. *Daily Mail*. Retrieved from http://www.dailymail.co.uk/news/article-3406830/I-m-forever-thankful-Jambo-gone-one-two-ways-Man-protected-huge-gorilla-fell-enclosure-boy-relives-dramatic-moment-30-years-says-t-wait-kids-there.html

107 Kidnapped girl 'rescued' by lions. (2005, June 22). *BBC*. Retrieved from http://news.bbc.co.uk/2/hi/africa/4116778.stm

108 Boyd, F. (2014, November 21). Experience: my horse saved me from a raging cow. *The Guardian*. Retrieved from https://www.theguardian.com/lifeandstyle/2014/nov/21/experience-my-horse-saved-me-from-raging-cow

109 People for the Ethical Treatment of Animals. *The Hidden Lives of Cows*. Retrieved from https://www.peta.org/issues/animals-used-for-food/factory-farming/cows/hidden-lives-cows/

110 Soomro, B.R. (2016, January 30). Remarkable animal rescues of humans. *Dawn*. Retrieved from https://www.dawn.com/news/1235930.

111 Chapman, P. (2004, February 18). Swimming cow saves farmer's wife. The Telegraph. Retrieved from https://www.telegraph.co.uk/news/worldnews/australiaandthepacific/newzealand/1454679/Swimming-cow-saves-farmers-wife.html

112 BBC News website. (2004, January 19). Rabbit saves diabetic from coma. Retrieved from http://news.bbc.co.uk/2/hi/uk_news/england/cambridgeshire/3441337.stm.

113 Morell, V. (2014, February 23). It's Time to accept that elephants, like us, are empathetic beings. *National Geographic*. Retrieved from https://news.nationalgeographic.com/news/2014/02/140221-elephants-poaching-empathy-grief-extinction-science/

114 Plotnik, J. (2014, February 18). Asian elephants (Elephas maximus) reassure others in distress. *PeerJ*. Retrieved from https://peerj.com/articles/278/

115 Holdrege, Craig (Spring 2001). Elephantine Intelligence. *In Context*. Retrieved 8 Jan 2019 from http://www.natureinstitute.org/pub/ic/ic5/elephant.htm

116 Mele, C. (2019, April 7). Rhino Poacher Killed by Elephant and Eaten by Lions, Officials Say. *The New York Times*. Retrieved from https://www.nytimes.com/2019/04/07/world/africa/south-africa-poacher-rhino-lions.html

117 Willett, W., et al. (2019, January 16). *Food in the Anthropocene: The EAT–Lancet Commission on healthy diets from sustainable food systems. The Lancet*. Retrieved from https://eatforum.org/initiatives/the-eat-lancet-commission/eat-lancet-commission-summary-report/

118 The statistics and references in this section were taken from www.undp.org

119 Food and Agircultural Organization of the United Nations. (2018). Transforming Food and Agriculture for the SDGs: 20 interconnected actions to guide decision-makers. Retrieved from http://www.fao.org/3/I9900EN/i9900en.pdf

120 Vinicius, M. (2011, June). Long-Lasting Effects of Undernutrition. *International Journal of Environmental Research and Public Health*. Retrieved from https://www.ncbi.nlm.nih.gov/pmc/articles/PMC3137999/

121 Eisler, P. Morrison, B. DeBarros, A. (2009, December 8). School Lunches Wouldn't Meet Fast-Food Safety Standards. *USA Today*. Retrieved from https://usatoday30.usatoday.com/news/education/2009-12-08-school-lunch-standards_N.htm

122 UNDP (2016) What are the Sustainable Development Goals? Retrieved from http://www.undp.org/content/undp/en/home/sustainable-development-goals.html

123 FAO, IFAD, UNICEF, WFP and WHO. (2017). The state of food and agriculture 2017: Leveraging food systems for inclusive rural transformation. *Food and Organization of the United Nations*. Retrieved from http://www.fao.org/3/a-i7658e.pdf

124 Food and Agricultural Organization of the United Nations. (2016). Sustainable Development Goals target 12.3 on food loss and waste: 2016 progress report. Retrieved from http://www.fao.org/save-food/news-and-multimedia/news/news-details/en/c/436985/

125 Thompson, P. & Kituyi, M. (2018, July 13). 90 percent of fish stocks are used up – fisheries subsidies must stop. *World Economic Forum*. Retrieved from https://www.weforum.org/agenda/2018/07/fish-stocks-are-used-up-fisheries-subsidies-must-stop/

126 Perlman, D. (2008, August 15). *Scientists alarmed by ocean dead-zone growth*. San Francisco Chronicle. Retrieved from https://www.sfgate.com/green/article/Scientists-alarmed-by-ocean-dead-zone-growth-3200041.php

127 Department of the Interior, U.S. Geological Survey. (2018, June 7) Press release: *Average-sized dead zone forecasted for the Gulf of Mexico*. Retrieved from https://www.usgs.gov/news/average-sized-dead-zone-forecasted-gulf-mexico

128 Neal, K. (2007, February 21). Global impact of livestock production focus of recent event. *Stanford News*. Retrieved from https://news.stanford.edu/news/2007/february21/livestock-022107.html

129 Brownell, C. (2016, September 29). The end of meat? Economics, the environment and changing tastes have top protein feeling the heat. *Financial Post*. Retrieved from https://business.financialpost.com/commodities/agriculture/the-end-of-meat-economics-the-environment-and-changing-tastes-have-top-protein-feeling-the-heat

130 Willett, W., et al. (2019, January 16). Food in the Anthropocene: the EAT–Lancet Commission on healthy diets from sustainable food systems. *The Lancet*. Retrieved from https://eatforum.org/initiatives/the-eat-lancet-commission/eat-lancet-commission-summary-report/

131 Harvard Medical School. (2012, January). What's the beef with meat? *Harvard Health Publishing*. Retrieved from https://www.health.harvard.edu/healthy-eating/whats-the-beef-with-meat

132 Plant Based Foods Association, Press Release, (2018, July 30). Plant Based Food Sales Grow 20 percent. Retrieved from https://plantbasedfoods.org/wp-content/uploads/2018/07/PBFA-Release-on-Nielsen-Data-7.30.18.pdf

133 Chiorando, M. (2018, August 1). US Plant-Based Meat Sales Hit $670 Million Skyrocketing By 24 percent. *Plant Based News*. Retrieved from https://www.plantbasednews.org/post/us-plant-based-meat-sales-hit-670-million-skyrocketing-by-24

134 Chiorando, M. (2018, July 4). Global Meat Substitutes Market To Exceed $6 Billion By 2023. Plant Based News. Retrieved from https://www.plantbasednews.org/post/global-meat-substitutes-market-to-exceed-6-billion-by-2023

135 Raphael, R. (2019, March 5). Beyond Meat just launched a plant-based product for beef lovers. *Fast Company*. Retrieved from https://www.fastcompany.com/90314807/beyond-meat-just-launched-a-plant-based-product-for-beef-lovers?partner=rss

136 Ibirogba, P. (2019, March 16). Vegan Company Beyond Meat's Plans to Lower Price Could Be Disastrous for Meat Industry. Vegan News. Retrieved from https://vegannews.co/vegan-company-beyond-meats-plans-to-lower-price-could-be-disastrous-for-meat-industry/

137 Brownell, C. (2016, September 29). The end of meat? Economics, the environment and changing tastes have top protein feeling the heat. *Financial Post*. Retrieved from https://business.financialpost.com/commodities/agriculture/the-end-of-meat-economics-the-environment-and-changing-tastes-have-top-protein-feeling-the-heat

138 Klementova, M.; Thieme, L.; Haluzik, M.; Pavlovicova, R.; Hill, M.; Pelikanova, T.; Kahleova, H. A Plant-Based Meal Increases Gastrointestinal Hormones and Satiety More Than an Energy- and

Macronutrient-Matched Processed-Meat Meal in T2D, Obese, and Healthy Men: A Three-Group Randomized Crossover Study. Nutrients 2019, 11, 157. Retrieved from https://www.mdpi.com/2072-6643/11/1/157

139 Murphy, P. (2019, April 19). 'Bug burger' idea wins UB World's Challenge Challenge. *UBNow*. Retrieved from http://www.buffalo.edu/ubnow/campus/campus-host-page.host.html/content/shared/university/news/ub-reporter-articles/stories/2019/04/worlds-challenge-challenge.detail.html

140 U.S. Food and Drug Administration (February, 2005). Retrieved from https://www.fda.gov/Food/GuidanceRegulation/GuidanceDocumentsRegulatoryInformation/SanitationTransportation/ucm056174.htm

141 Tousignant, L. (2017, August 16). Bug 'burgers' are hitting supermarket shelves. New York Post. Retrieved from https://nypost.com/2017/08/16/bug-burgers-are-hitting-supermarket-shelves/

142 Schaefer, O. G. (2018, September 14). Lab-Grown Meat: Meat produced without killing animals is heading to your dinner table. *Scientific American*. Retrieved from https://www.scientificamerican.com/article/lab-grown-meat/

143 Okin GS (2017) Environmental impacts of food consumption by dogs and cats. *PLoS ONE 12(8): e0181301*. Retrieved 2019, January 13 from https://doi.org/10.1371/journal.pone.0181301

144 Association of American Feed Control Officials. What is in Pet Food? Retrieved from https://www.aafco.org/consumers/what-is-in-pet-food

145 Springmann M, Mason-D'Croz D, Robinson S, Wiebe K, Godfray HCJ, Rayner M, et al. (2018, November 6). Health-motivated taxes on red and processed meat: A modelling study on optimal tax levels and associated health impacts. *PLoS ONE* 13(11): e0204139. Retrieved from https://doi.org/10.1371/journal.pone.0204139

146 The North American Meat Institute. (2018). The United States Meat Industry at a Glance. Retrieved from https://www.meatinstitute.org/index.php?ht=d/sp/i/47465/pid/47465

147 UN Food and Agriculture Organization. (2018). Decent Rural Employment. Retrieved from http://www.fao.org/rural-employment/agricultural-sub-sectors/livestock/en/

148 Vegetarian Times Editors (2014, November 26). How to buy veg-friendly cheese. *Vegetarian Times*. Retrieved from https://www.vegetariantimes.com/skills/how-to-buy-veg-friendly-cheese

Made in United States
North Haven, CT
10 November 2026

22091252R00097